Media for Business

by Robert H. Amend
and Michael A. Schrader

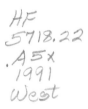

HF
5718.22
.A5x
1991
West

Knowledge Industry Publications, Inc.
White Plains, New York

The Video Bookshelf

MEDIA FOR BUSINESS

Robert H. Amend
Michael A. Schrader

ISBN 0-86729-2644

Printed in the United States of America

Copyright © 1991 Knowledge Industry Publications, Inc.
701 Westchester Avenue, White Plains, NY 10604

Not to be reproduced in any form whatever without written permission from the publisher.

10   9   8   7   6   5   4   3   2   1

Detailed Table of Contents

List of Figures and Tables xi
Introduction xiii

1. **Effective Presentation Planning** 1
   The Presentation Analysis 1
      Background 2
      Audience 2
      Goals 3
      Administrative Information 3
   Using the Presentation Analysis Form 7

2. **Budgeting** 9
   Budgeting Environments 9
   Cost Centers 10
   Overhead Systems 11
   Hybrid Systems 11
   Using a Budget Form 12
      Labor 12
      Purchased Services 16
      Rentals 18
      Supplies 18
      Totals 21

3. **Scripting** 23
   The Scripting Process 23
      Research 24
      The Treatment 24
      The Rough Draft 25
         The Split Page 25
         The Storyboard 27
         The Corporate Teleplay 27
   Script Elements 27
   Script Evaluation 31

   Script Payment Methods 31
   Summary 32

4. **Audio** 33
   The Elements of Audio 33
     Voice 34
     Music 35
       Original Music 37
       Using Copyrighted Music 37
       Music Libraries 38
     Other Audio Elements 39
   Audio Equipment 41
     Recorders 42
     Playback Equipment 43
     Microphones 43
     Other Equipment 45
   Audio Studios 47
   Stand-alone Audio Programs 47
   Summary 48

5. **Using Graphics** 49
   Types of Business Graphics 49
     Charts and Graphs 51
       Charts and Graphs Using an X-Y Axis 51
       Pie Charts 52
       Tables 53
     Illustrations 53
     Cartoons 53
     Animation 55
   Common Graphics Pitfalls 55
     Conveying Too Much Information 56
     Using Color Improperly 56
     Ignoring the "Safe Title" Area 56
   Producing Business Graphics 57
     Desktop Publishing Systems 57
     Computer Graphics 58
     Animation 59
   Summary 59

6. **Print for Media Presentations** 61
   Types of Printed Materials 62
     Handouts 62

	Pamphlets 62

        Pamphlets 62
        Briefing Books 62
        Brochures 63
    Deciding When to Use Print to Support a Media Program 63
        Example: Printed Material to Support a Fundraising
            Project 63
        Example: Printed Material to Supplement an
            Employee Indoctrination Program 64
        Example: Printed Materials to Support an Overhead
            Transparency or Slide Presentation 65
        Example: Printed Materials to Support a Multi-Image
            Program 66
    Conclusion 67

7. **Speaker-Supported Presentations** 69
    Advantages of Speaker-Supported Presentations 70
        Minimal Production Time 70
        Minimal Production Costs 71
        Presentation Flexibility 71
        Audience Interaction 72
        Facts and Figures 72
    Disadvantages of Speaker-Supported Presentations 72
        Inconsistency 73
        Diminished Impact 73
        Non-Entertaining 74
    Producing a Speaker-Supported Presentation 75
        Production 75
            Media Producer 76
            Artist 76
            Typesetter 76
            Photographer 76
            Graphic Arts Specialist 77
        Production Options 77
            Example: Overhead Transparencies 77
            Example: A More Professional Look 79
            Example: Color Visuals 80
                Reflected Copy 80
                Computer Graphics 81
        Desktop Publishing 82
    Presentation Equipment and Techniques 83
        Showing Overhead Transparencies 83
        Showing 35mm Slides 83

v

       Computer-Assisted Presentations 86

8. **Slide/Cassette Programs** 87
   - Producing a Slide/Cassette Program 88
     - Planning 88
     - Budgeting 89
       - Script Fees 89
       - Film Costs and Photography Fees 90
       - Audio and Graphics 90
       - Equipment 91
       - Supplies and Miscellaneous Equipment 91
     - Scriptwriting 92
   - Production Planning 92
     - Photography 93
     - Graphics 94
     - Audio 96
     - Assembling and Programming the Show 97
     - Presentation Techniques and Equipment 98

9. **Multi-Image** 99
   - Pre-Production Considerations 100
     - Planning 101
     - Budgeting 103
     - The Script 105
     - The Photographer, Film and Processing 105
     - Stock Photos and Graphics 106
     - Audio 107
     - Programming 107
     - Contingency Items 108
   - The Production Process 108
     - Scriptwriting 108
     - Photography 109
     - Audio 111
     - Staging 111
     - Transfer to Videotape 113
   - Conclusions 114

10. **Film Production** 115
    - Film to Video: A Short History 116

	Advantages of Film 116
		Appearance 116
		Presentation 117
		Duplication 117
	Disadvantages of Film 118
		Reliance on Vendor Services 118
		Cost 118
		Turnaround Time 119
Producing a Film 119
	Pre-production 119
		The Presentation Analysis 120
		Developing the Script: Research to Treatment
			to Script 120
		Selecting a Production Crew 121
		Equipment 122
		Selecting Performance Talent 123
		Scouting Locations 124
	Production 125
		The Crew 125
			Director 125
			Cameraperson 126
			Audio Technician 126
			Production Assistants 126
			Grips 127
		Setting Up the Shot 127
		Production Administration 128
	Post-production 129
		Editing 133
		The Audio Mix 134
		Conforming 136
The Future of Film in Business and Industry 137

11. **Videotape Production** 139
	Planning a Videotape Production 141
		Presentation Techniques 141
		Budgeting 142
		Remote versus Studio Production 143
	Pre-Production 144
	Production 146
		Visuals 146
		Sound 147

	Post-production	148
	Formats and Equipment	149
	The Various Formats	150
	The Equipment	151
	Cameras	152
	Recorders	153
	Lights	153
	Microphones	155
	Other Equipment	156
	Summary	161
12.	**Interactive Video**	163
	Weighing the Differences	164
	Advantages of Interactive Video	164
	Disadvantages of Interactive Video	166
	Interactive Formats	167
	Levels of Control	169
	Budget Considerations	169
	Personnel	169
	Equipment	170
	Production Design Considerations	171
	Applications	175
	Summary	177
13.	**Teleconferencing**	179
	Introduction	179
	Audioconferencing	184
	Introduction	184
	Applications	184
	Extra Sales Revenue	185
	Contract Negotiations and Reviews	185
	Continuing Medical Education	185
	Focus Groups	186
	Group Discussion	186
	Distance Education	186
	Tips for Planning an Audioconference	187
	Interconnection Options.	187
	Scheduling Arrangements	188
	Conclusion: Preparing for the Audioconference	189
	Audiographic Conferencing	190
	Introduction	190

Audiographic Conferencing Systems 190
Applications 191
    Educational Institutions 192
    Corporations 192
    Hospitals 193
    The Military 193
Benefits and Limitations of PC-based Audiographic
    Conferencing 193
Videoconferencing 195
    Introduction 195
    Corporate Videoconferencing Applications 195
        Banking 196
        Retailing 196
        Engineering 197
        Telecommunications 197
        Insurance 197
    Medical Videoconferencing Applications 197
    Government Videoconferencing Applications 198
        U.S. Army Materiel Command 198
        U.S. Naval Underwater Systems Command 198
        National Aeronautics and Space Administration 199
        Local Justice Departments 199
    Education Videoconferencing Applications 199
        Pennsylvania State University 199
        University of Missouri 199
    Videoconferencing Technology 200
        Video Compression 200
    Videoconferencing Networks 201
    Videoconferencing Systems 202
    Connecting the Systems 202
    Multipoint Videoconferencing 204
Business Television 204
    Introduction 204
    Benefits of Business Television 207
        Immediacy 207
        Simultaneous Message Delivery 207
        Feedback from the Field 208
        Greater Access to Experts 208
        Efficiency of Training 208
        Travel Reduction 208
        Motivation 209
        Increased Productivity 209

	Business Television Applications	209
	Training	209
	Employee News/Information	210
	Product Announcements	210
	Press Conferences	211
	Special Events	211
	Cross-Networking	211
	External Programming	212
	Conclusion	212
	Further Information	213
	Periodicals	213
	Books	214
	Organizations and Conferences	214

14. **Evaluating Media** 215
    Developing Evaluation Techniques 215
        A Continuum 216
        A Model 217
    Conclusion 221

Glossary 223
Bibliography 237
About the Authors 241

List of Figures and Tables

Figure 1.1:	Presentation analysis	4
Figure 2.1:	Sample budget form	13
Figure 2.2:	Labor	16
Figure 2.3:	Purchased services	17
Figure 2.4:	Rentals	19
Figure 2.5:	Supplies	20
Figure 2.6:	Totals	20
Figure 3.1:	The split page script	26
Figure 3.2:	A storyboard	28
Figure 3.3:	The corporate teleplay	29
Figure 4.1:	A narrator's script	36
Figure 4.2:	An E-MAX sampling device	40
Figure 4.3:	4-track reel-to-reel tape deck	42
Figure 4.4:	Basic microphone patterns	44
Figure 4.5:	Examples of microphones	45
Figure 4.6:	A 6 x 2 audio mixer	46
Figure 5.1:	Bar chart, example of X-Y axis chart	50
Figure 5.2:	Example of a pie chart	51
Figure 5.3:	Example of a table	52
Figure 5.4:	Example of an illustration	54
Figure 7.1:	3M transparency maker	78
Figure 7.2:	Genegraphics computer graphics station	81
Figure 7.3:	Bell & Howell overhead transparency projector	84
Figure 7.4:	Briefcase overhead projector; open and closed	84
Figure 7.5:	Loading carousel onto a 35mm slide projector	85
Figure 8.1:	Kroy type machine	95
Figure 8.2:	Photo copystand	95

Figure 8.3:	GPG turnkey computer graphics system	96
Figure 9.1:	Screen configuration examples	102
Figure 9.2:	Multi-image equipment: Projector, reel-to-reel recorder/player and programmer	104
Figure 9.3:	Using a Bencher photo copy stand	110
Figure 10.1:	Camera report	130
Figure 10.2:	16mm full coat	131
Figure 10.3:	Log sheet	132
Figure 10.4:	Edge number on 16mm film	133
Figure 11.1:	A switcher/special effects generator	144
Figure 11.2:	A shot list	145
Figure 11.3:	Videotape edit controller	149
Table 11.1:	Comparison of selected tape formats	150
Figure 11.4:	Videotape field recorder	153
Figure 11.5:	3-point or triangular lighting	154
Figure 11.6:	A wireless, lavalier microphone	155
Figure 11.7:	ENG fluid head tripod	157
Figure 11.8:	Dolly, truck and arc camera movements	158
Figure 11.9:	Teleprompter (front) mounted on a video camera	159
Figure 11.10:	Videotape edit system	159
Figure 11.11:	A character generator	160
Figure 12.1:	Laser videodisc player	168
Figure 12.2:	Sample flowchart—"The Art and Science of Making Bread"	173
Figure 12.3:	Simple branching	174
Figure 13.1:	A Matrix of Teleconferencing Technologies	182
Figure 13.2:	Business Television Networks and Services	205
Figure 14.1:	Range from qualitative to quantitative techniques	216
Figure 14.2:	Formative evaluation	220

INTRODUCTION

More than ever before, good communication skills are a necessity in the business world. Simply doing a good job is not enough. People in business today must also be able to communicate their accomplishments to others in a professional manner. This requires both solid communications skills and a thorough understanding of the media commonly used in business and industry today.

This book will introduce you to the various types of media used in the business world. It will provide you with a basic understanding of how media presentations are produced so you can apply this information to your business communication efforts. We have designed this book so that it will be useful to a wide range of professionals. Managers, engineers, trainers, salespeople and others who participate in or supervise the production of a business program will find this book helpful. Beginning and intermediate students involved in media production will gain valuable insight into the world of business communication.

The book is divided into two sections. The first six chapters address the various facets of media production: presentation planning, budgeting, scripting, audio, and preparation of graphics and printed support materials. These are critical to any media production—regardless of format—and have direct impact on all media covered in this book. Refer to these chapters before you select a specific media format.

Chapters 7-13 address specific media formats commonly used in business and industry: speaker support, tape/slide, multi-image, film, videotape, interactive video, and teleconferencing. The novice communicator will receive a basic, working understanding of each, as these chapters discuss the advantages and disadvantages of each medium, specific media production techniques, budget considerations and other pertinent issues.

Our concluding chapter, "Evaluating Media," will provide you with some guidelines for measuring and understanding the effectiveness of media programs.

It is our hope that after reading this book, you will have a better understanding of the media tools available to today's business communicator. We also hope that you are able to use this book as a reference for your future communication efforts.

1
Effective Presentation Planning

Going into a manager's meeting on union relations, we were prepared for a long-winded lecture filled with legal jargon and irrelevant case studies. We were wrong. The presentation was informative, entertaining, and full of information so relevant to our daily activities we felt the speaker had prepared his presentation especially for us. The 200 other managers in the audience clearly concurred and we showed our appreciation with a hearty round of applause. Later, many attendees offered their personal thanks as well. We wondered what had set this presentation apart from so many others we had attended, and took our question to the speaker. After a short conversation, the answer became clear. The difference was planning.

This presentation had not been a generic "off-the-shelf" speech delivered to general audiences. Nor had it been a hastily prepared collection of handwritten charts and graphs. Prior to stepping up to the podium, this speaker had given his presentation a great deal of thought. He was not only an authority on the subject matter, he had gone beyond content and researched the presentation itself. He had learned about his audience—who we were, why we were there, where we worked. He'd looked into union relations at our facility and tailored his presentation to address our specific needs. He had set specific goals that his presentation would try to achieve. And finally, he had looked into the presentation site in advance to ensure that all would be ready when he arrived. In short, our speaker had done his homework.

THE PRESENTATION ANALYSIS

In business terms, this kind of "homework" is called a presentation analysis. It is a structured approach to collecting information about a presentation prior to its production.

The presentation analysis is critical to planning an effective program or event because the information collected will help determine the presentation's content, direction, style and visual media. To perform such an analysis, you will need information about the following: background, audience, goals and administrative details.

The presentation analysis form included here (see Figure 1.1) is an example of one we have used successfully in the past. Let's examine the four categories it covers.

Background

This section of the analysis addresses presentation content. It should serve to identify the broad subject you will cover as well as specific points within the overall subject that must be stressed or reinforced. Often, after completing the background portion of the presentation analysis, the client realizes that the original concept for the program must be enlarged or scaled down. For example, if the client lists 10 or 15 subjects that must be covered in the presentation, it will be evident that several programs, not one, will actually be required.

You will need to know why a presentation is required; are you addressing a problem, proposing a solution, introducing a new concept? Are there any sensitive or political factors that might affect how the subject must be handled? Are there any problems or misunderstandings that must be addressed or corrected?

The background portion of the questionnaire is not a substitute for content research. Completing it will provide you with valuable insight into the specific aspects of your subject that must be covered while highlighting any specialized treatments that may be required.

Oddly enough, few presenters pay enough attention to this area. Most presenters are familiar with their subject matter and know exactly what they want to tell their audience. What they often fail to consider, however, is what their audience wants or needs to know about the subject. Knowing your audience can make the difference between a successful presentation and one that falls short of the mark.

Audience

This section is designed to introduce you to the people in your audience in order to help you tailor your presentation to their needs. How many people will you be addressing? How old are they? Where do they work? Will you be addressing men, women or both? What is their educational background? Any additional demographic information you can obtain will help you design your presentation more effectively.

You will also need to determine what your audience already knows about the subject. Do they already have positive or negative attitudes towards the subject? Be sure to consider why the audience will be watching your program. Is attendance optional or will they be required to attend? What will the audience gain by watching the presentation? You should also examine the possibility of a secondary audience. Who else might benefit from viewing this presentation? Secondary audiences also need to be researched, but not to the extent of the primary one.

Goals

Every presentation should have a well-stated goal or purpose. What do you hope to accomplish? Are you trying to inform, motivate or persuade? Are you trying to teach or train? For instance, if you are producing a program encouraging auto salesmen to sell more cars, the purpose of your program would be to motivate. On the other hand, if your program is intended to tell the same audience about the latest requirements for filling out a sales proposal, your goal would be to inform.

The goal of your presentation must be defined as specifically and in as much detail as possible. A program that tries to be all things for all audiences is doomed to mediocrity. A well-defined goal statement is invaluable when determining the appropriate style, content and visual media for your presentation. For example, suppose a client approaches you to produce a videotape teaching machinists how to operate a specialized drill press. Naturally, you assume that the purpose of the program is training. While completing the presentation analysis, however, you learn that several operators have recently injured themselves while using the machine. The real goal of the program—promoting safety—becomes apparent. This discovery changes the type of training the program must provide. Now the program will focus on safety procedures as well as basic operating techniques.

Administrative Information

This section of the questionnaire provides you with general information you will need to know in order to administer your presentation effectively.

Figure 1.1: Presentation analysis

PROJECT ANALYSIS

Project Name _____
Client/Customer _____ Phone # _____
Producer/Director _____ Phone # _____
Writer _____ Phone # _____

BACKGROUND

1. Describe the situation prompting you to request an audiovisual production.

2. Which of these subjects does the project deal with? (Choose 1 or more)

3. Rank, in order of importance, the major subjects that must be addressed.

4. What specific objectives does this program address?

5. Are there any particular problems or misunderstandings about the subject that must be corrected?

6. Are there any political taboos or political factors we must consider?

Figure 1.1: Presentation analysis (cont.)

AUDIENCE

1. Describe the primary audience (include total number, job descriptions and levels, job locations, sex, age range, educational background, etc.).

2. What is their previous background, if any, on the subject?

3. Why should the audience watch this program? List the major benefits.

4. Does the audience have any negative attitudes towards the subject? Why?

5. What knowledge or attitudes should the audience have after viewing the program?

6. Is there a secondary audience? Please describe.

Figure 1.1: Presentation analysis (cont.)

GOALS

1. What do you want this program to do? (Explain briefly and check no more than two)

   A. Inform
   B. Motivate
   C. Teach
   D. Change an attitude
   E. Other

2. What specific skills, perceptions or attitudes should be developed in the audience?

3. In one sentence, recap the most important idea the audience should perceive after viewing this program?

ADMINISTRATIVE INFORMATION

1. What is the target date for completion?

2. What is the reason for this date?

3. Who can change the date if necessary?

4. In how many locations will the program be shown?

   A. In what kind of situations?

      — small groups
      — large groups
      — one-on-one
      — supervised
      — unsupervised

5. How long will the program remain current?

Figure 1.1: Presentation analysis (cont.)

> 6. What is your budget for production?
>
> 7. Who can amend the budget if necessary?
>
> 8. Who will be assigned as technical expert? (include phone number)
>
> 9. Are there any presentational styles that would be inappropriate?
>
> 10. Are there any existing visual materials or similar programs on the subject?

There are obvious items such as date, time and location. But, you will also need to know the amount of funding available for production elements, such as scripting and photography. If you need technical experts you will need to identify and locate these people. You will need to determine if the production requires research material, reports or photographs from another location.

In this section of the analysis you can identify existing visual materials on the subject. You may want to review these materials either for use in your presentation (with permission of the originator) or for reference in your research.

It is in this part of the analysis that you will want to make note of any production constraints. Often the nature of a presentation requires or precludes certain production styles or techniques. For example, it would be unwise to produce a videotape for use at a facility equipped with only a 35mm slide projector.

Using the Presentation Analysis Form

For maximum effectiveness, the presentation analysis form should be completed during a meeting between the program producer and the program requestor. Participants at this meeting might include a client and a professional media producer, or a manager and an employee assigned to produce a presentation. In any case, the person producing the presentation should complete the form based on information provided by the requestor. Also present at the meeting should be any other individuals who have a say in the presentation so any misunderstandings can be worked out before they become a problem. Once a program is in production it is very time-consuming and costly to make changes that should have been worked on earlier.

Keep in mind that the presentation analysis form can and should be modified when necessary to fit your specific requirements. Some questions may not be applicable to your needs; other questions may need to be added or changed. The basic categories will probably remain the same.

Upon completion of the form you should have an excellent feel for the presentation you are about to produce. You should be able to identify your subject and the background information related to it. You should have a better understanding of your audience and what they expect from your program. You should also have established a well-defined goal your presentation is to achieve. Finally, you will have determined significant administrative factors related to your program such as date, time, location and the level of funding available for production costs.

Your next challenge is to determine the appropriate visual media for your presentation. The choices are many, ranging from the simple and inexpensive overhead transparency to the more complex and costly videotape or film production. As we examine the types of media most commonly used in business and industry today, you'll see how the information gathered in the presentation analysis will help you to choose the right media for your presentation. But first, a look at the bottom line—budgeting—is in order.

2
Budgeting

The budget for a media program is determined by a variety of factors, among them the intended use of the program, the level of production value required, the budgetary system used by the company producing the program, and the number of dollars available for media production. In this chapter we'll examine how corporate budgets are structured and what effect this has on the media producer. We'll also present a budgeting form that can be used for the media most commonly used by business and industry today. (It is important to note that although the budget information presented in this chapter is primarily designed for media producers working within a large corporation, it will also be valuable to those working with small scale business and industrial clients.)

Establishing a budget for a media program is the first major step in the production process. Funding levels have an impact on nearly all later program decisions. Will the program consist of a simple slide/tape production or a video extravaganza complete with professional talent and original music? Will the program be produced in-house or will an outside production company be hired to do the job? The budget has to be set before these and many other production questions can be answered.

Remember that the amount of money in the budget shouldn't be an immediate determinant of how much the production will cost. It is often possible to produce a very acceptable program without spending all of the available funds. The budget, like all other production factors, must be matched to the needs of the program. A budget that is too large may result in a final program that is full of all the latest special effects, but so overpowers the audience that it fails to make its point. On the other hand, an undersized budget may force the producer to cut corners to such an extent that the success of the program is compromised.

BUDGET ENVIRONMENTS

Business accountants employ a wide variety of bookkeeping methods, each of which tracks expenditures differently. Regardless of the system, however, there are some similarities and common categories. For our purposes, we'll keep the descriptions simple and focus on how budgetary systems affect the media production process.

Most business and industrial media production groups utilize one of three basic budgeting techniques: cost center (chargeback), overhead or hybrid. Each system places different requirements on the media producer who is preparing program budgets. Following is a rundown of how each works.

Cost Centers

This budgetary system, also known as chargeback, demands the most involvement from the producer since it requires the media production department to budget for the program as would an outside agency contracted for a specific job. Under this system, the in-house media production department is like a business within a business: the customers or departments requesting the media package ultimately pay all costs associated with the program. This includes all salaries, purchased services (such as film processing and post-production support) and supplies or equipment rentals. Therefore, a very detailed budget must be prepared as soon as the project analysis has been completed and approved.

This budget, which should include all costs the producer expects will be associated with the project, is submitted to the client for approval before any production steps are initiated. At this point the producer should be prepared to answer the client's questions on any and all aspects of the budget. This includes justifying the number of hours planned for production, the salaries of freelance talent and the rental costs for top-of-the-line production equipment.

If funding is tight and the client has reservations about the project costs, the producer and the client will have to negotiate a compromise that meets the needs of both. An amateur narrator may be substituted for a professional, or expensive special effects may be dropped in favor of less costly word charts. Once the final budget figure has been negotiated and approved, the production process can begin.

Cost centers provide a challenging fiscal environment for corporate media producers. Like freelancers, media producers working within cost centers depend on clients for their livelihoods. They have to walk the line between providing reasonably priced in-house media production services while generating enough funds to support their ongoing efforts. Because cost centers typically receive minimal corporate financial support, media departments operating within these systems must generally turn at least a modest profit in order to survive.

Overhead Systems

In an overhead budget environment, the media department is given an operating budget of its own and does not charge its clients for media production. Personnel salaries, purchased services, supplies, and any rental charges are paid out of the departmental budget. Procedurally, in many ways this system is the opposite of a cost center in that it treats the media department as an overhead expense. From a bookkeeping standpoint, the media department is treated much the same as the personnel, purchasing, mail, and other departments working in support of the overall company mission. The media department still receives work from its in-house clients; however, it does not depend on clients for financial resources since all production costs are paid out of the media production department's budget.

The primary difference between planning a program within an overhead system versus a cost center system is that the media department—not the client—determines the program's budget. Because labor, purchased services, equipment rentals and the like are paid for by the production department, it is important for the department to monitor the allocation of its resources on a continual basis. Current project demands must be balanced against future need in order to protect the department's financial health. Overhead systems are easier to administer than are chargeback systems, since all expenses are paid from a single operating budget. In fact, formal cost breakdowns may not even be required for smaller productions such as speaker-supported presentations or modifications to existing programs.

Hybrid Systems

Hybrid budgetary systems incorporate elements of both cost centers and overhead systems. Media departments operating under a hybrid system are allocated an operating budget for certain departmental expenses which may include labor, administrative supplies and travel expenses. Other production expenses such as hiring of professional on-camera talent, purchasing special effects and equipment rentals are charged to the client's budget. There is no standard for hybrid systems and they vary from company to company. If you work in a hybrid system, don't forget to carefully negotiate production costs with your clients so that each party is clear on who is responsible for each.

Using a Budget Form

The budget form presented here (see Figure 2.1) is designed to help you plan and record media production expenditures. The form covers the following four basic budget categories: labor, purchased services, rentals and supplies. Labor generally includes the salaries for the director, the actors, the cameraman and any other personnel working on the program. The category of purchased services covers all services purchased in support of the production, such as film processing, travel, music recording, casting and sound mixing. The category of rentals accounts for any equipment or facilities rented in support of a production, and supplies covers film, videotape, paper, pencils and the like.

You will note that each section of the form is totaled individually and that these subtotals are indicated on the last page of the form, along with the overall budget. The form includes common expenses found within each category. For example, the Labor section lists the following positions: director, producer, writer, three photographers, two production assistants, three talent positions and a graphic artist. These are common positions found on many media productions. Naturally, each position will not be filled for each production. Blank lines are included under each category for expenses not listed in the standard form. These forms can be modified and reproduced to fit any special requirements you have.

Let's look at a completed form for a modest videotape production and examine each entry for each budget category. Remember that costs vary widely from area to area and are not necessarily the same as those you might encounter where you do business.

Assume that we're producing an industrial videotape for a corporate client. The purpose of the program will be to train employees how to use an expensive new drill press the company is purchasing. The client has indicated that the training is very important to the company's goals and that professional on-camera talent is a must. We'll also be renting production equipment and contracting with freelance talent to fill key production positions. The program will be shot both on location and in the studio. Production will take one week and the client has budgeted $22,000 for this program.

Labor

The first entry in Figure 2.2 shows $3000, the producer's fee for this program. Our production crew will consist of a director, a videographer (listed

Figure 2.1: Sample budget form.

```
                        Project
                     Budget Sheet

Project Title_____  Date _____
Producer_____ Client _____

SECTION I—LABOR

Personnel                         Cost/Hr.    Hours    Cost

 1. Producer                      _____    _____  _____
 2. Director                      _____    _____  _____
 3. Writer                        _____    _____  _____
 4. Photographer I                _____    _____  _____
 5. Photographer II               _____    _____  _____
 6. Photographer III              _____    _____  _____
 7. Production assistant I        _____    _____  _____
 8. Production assistant II       _____    _____  _____
 9. Talent I                      _____    _____  _____
10. Talent II                     _____    _____  _____
11. Talent III                    _____    _____  _____
12. Graphic artist                _____    _____  _____
13. _____               _____    _____  _____
14. _____               _____    _____  _____
                                  Total Labor Costs_____

SECTION II—PURCHASED SERVICES

Service                           Cost

 1. Scripting                     _____    _____  _____
 2. Casting                       _____    _____  _____
 3. Narration recording           _____    _____  _____
 4. Music (library)               _____    _____  _____
 5. Music (original composition)  _____    _____  _____
 6. Post Production Services (Video) _____ _____  _____
    a. Editing                    _____    _____  _____
    b. Dubbing                    _____    _____  _____
    c. Special FX                 _____    _____  _____
    d. Animation                  _____    _____  _____
    e. _____            _____    _____  _____
    f. _____            _____    _____  _____
```

Figure 2.1: Sample budget form (cont.)

SECTION II—PURCHASED SERVICES (cont.)

Service   Cost

7. Motion Picture Services
   a. Film processing   ____
   b. Work printing   ____
   c. Audio transfers   ____
   d. Edge numbering   ____
   e. Answer prints   ____
   f. Release prints   ____
   g. Sound mix   ____
   h. _____   ____
   i. _____   ____
8. Still Photographic Services
   a. Film processing   ____
   b. Slide mounting   ____
   c. Printing   ____
   d. Special fx   ____
   f. Slide duplication   ____
   g. _____   ____
   h. _____   ____
9. Catering _____   ____

                Total Purchasing Services _____

SECTION III—RENTALS

Item   Cost

1. Facilities
   a. Studio   ____
   b. Location I   ____
   c. Location II   ____
   d. Location III   ____
   e. _____   ____
   f. _____   ____
2. Equipment
   a. Video camera(s)   ____
   b. Motion picture camera(s)   ____
   c. Still photographic camera(s)   ____
   d. Tripod(s)   ____
   e. Camera dolly(s)   ____

Figure 2.1: Sample budget form {cont.}

```
SECTION III—RENTALS (Cont.)

Item                                    Cost

    f.  Aerial camera mount(s)          ____
    g.  Lenses                          ____
    h.  Batteries                       ____
    i.  Lighting kit(s)                 ____
    j.  _____              ____
    k.  _____              ____
    l.  _____              ____
 2. Props
    a.  _____              ____
    b.  _____              ____
    c.  _____              ____
 3. Wardrobe/Costumes
    a.  _____              ____
    b.  _____              ____
    c.  _____              ____
                    Total Rentals   _____

SECTION IV—SUPPLIES

Item                                    Cost

 1. Videotape                           ____
 2. Motion picture film                 ____
 3. Still photographic film             ____
 4. Misc. administrative supplies       ____
 5. _____          ____
 6. _____          ____
                    Total Supplies _____

               TOTAL COSTS
• • • • • • • • • • • • • • • • • • • • • • • • • •
                         Total Labor  _____
              Total Purchased Services _____
                       Total Rentals  _____
                      Total Supplies  _____
                        Grand Total   _____
```

Figure 2.2: Labor

Personnel	Cost/Hr.	Hours	Cost
1. Producer			$3000.00
2. Director	$35.00	80	2800.00
3. Writer			
4. Photographer I	20.00	40	800.00
5. Photographer II			
6. Photographer III			
7. Production Assistant I	10.00	40	400.00
8. Production Assistant II			
9. Talent I	—	—	1200.00
10. Talent II			
11. Talent III			
12. Graphic artist	15.00	8	120.00
13. _____			
14. _____			
	Total Labor Costs		$8320.00

under photographer, this person will operate the video camera), and a production assistant. The director must be present for both the production and editing so this person will be needed for a full two weeks (80 hours). Because the photographer and the production assistant will be required only for the field and studio production, their services are needed for just one week (40 hours). The on-camera narrator (Talent 1) will be needed for three days of shooting. The fee for this service is $400 per day, or $1200 dollars in all. We will also need an artist to design several graphics for the program. This will take one day and will cost $120. The total labor budget for this program will be $8320.

Purchased Services

The next budget category is Purchased Services. The completed section is shown in Figure 2.3.

The script for this program is being produced by a freelance scriptwriter and will be completed prior to production. The fee for this script is $2000. A

Figure 2.3: Purchased services

Service	Cost
1. Scripting	$2000.00
2. Casting	350.00
3. Narration recording	200.00
4. Music (library)	500.00
5. Music (original composition)	_____
6. Post production services (video)	
a. Editing	3200.00
b. Dubbing	
c. Special FX	_____
d. Animation	1000.00
e. _____	_____
f. _____	_____
7. Motion picture services	
a. Film processing	_____
b. Work printing	_____
c. Audio transfers	_____
d. Edge numbering	_____
e. Answer prints	_____
f. Release prints	_____
g. Sound mix	_____
h. _____	_____
i. _____	_____
8. Still photographic services	
a. Film processing	_____
b. Slide mounting	_____
c. Printing	_____
d. Special FX	_____
e. Slide duplication	_____
f. _____	_____
g. _____	_____
9. Catering _____	_____
Total Purchased Services	**$7250.00**

casting session will be required to select the individual who will be our on-camera talent. The fee for this casting is $350. Any narration which is not delivered on-camera will be recorded at a sound studio. Recording time runs about $100/hour and we have allowed for two hours of recording time. Music for the program will be selected from a music library. We have budgeted $500 for royalty fees and re-recording of our selections. Editing will take place at a local post-production facility. The fee for use of the facilities and an operator is $200/hour. We have allowed for two full days of editing at a cost of $3200. To illustrate several concepts not easily photographed, we have planned for some very simple computer animation to be included in the program and have budgeted $1000 for this purpose. The total budget for purchased services is $7250.

Rentals

The next budget category on the form is rentals. The completed section is shown in Figure 2.4.

Production techniques used in this program will be straightforward and will not require a great deal of equipment. Two days of shooting will require a studio, that rents for $350 per day. Video camera rental is $300/day for five days. Additional camera equipment will include video recorders, a tripod, batteries, and a lighting kit. The costs for these items are listed on the sample section. The total rental budget is $3320. (Many additional camera accessories such as light meters, filters and gels, are provided by the photographer and do not need to be rented for this production.)

Supplies

The final section of the budget form accounts for supplies (see Figure 2.5). Very few supplies are needed for this production. The most significant expense under this category will be for videotape. We have budgeted $500 for 20, 3/4-inch field production tapes (at $15 each) and approximately 12 tapes for editing and duplication. We have also budgeted $100 for administrative supplies such as paper and pencils. An emergency fund of $300 has been set aside to replace burned-out lamps in the lighting kit and to purchase electrical supplies (i.e., connectors or adaptors) needed during production. The total supplies budget is $900.

Figure 2.4: Rentals

Item	Cost
1. Facilities	
a. Studio	$700.00
b. Location I	_____
c. Location II	_____
d. Location III	_____
e. _____	_____
f. _____	_____
2. Equipment	
a. Video camera(s)	1500.00
b. Motion picture camera(s)	_____
c. Still photographic camera(s)	_____
d. Tripod(s)	120.00
e. Camera dolly(s)	_____
f. Aerial camera mount(s)	_____
g. Lenses	_____
h. Batteries and chargers	600.00
i. Lighting kit(s)	500.00
j. _____	_____
k. Field recorders	600.00
l. _____	_____
2. Props	
a. _____	_____
b. _____	_____
c. _____	_____
3. Wardrobe/Costumes	
a. _____	_____
b. _____	_____
c. _____	_____
	Total Rentals $4020.00

Figure 2.5: Supplies

Item	Cost
1. Videotape	$500.00
2. Motion picture film	_____
3. Still Photographic Film	_____
4. Misc. administrative supplies	100.00
5. _____	_____
6. Electrical supplies	300.00
Total Supplies	$900.00

Figure 2.6: Totals

TOTAL COSTS	
Total Labor	$8320.00
Total Purchased Services	7250.00
Total Rentals	4020.00
Total Supplies	900.00
Grand Total	$20,490.00

Totals

The final page of the budget (Figure 2.6) shows the grand total for all expenses. The program is budgeted at $20,490, $1510 less than the client's total of $22,000. The balance will be used as a contingency fund for unanticipated expenses. Unused funds will be returned to the client.

Naturally, this budget form will not meet all of the requirements for every media production. However, with some modification, this form and the information contained in this chapter, should help form a solid foundation for your future budgeting efforts.

The next step in the process in preparing a script.

3
Scripting

Let's introduce you to a supervisor and a member of her staff, neither of whom are media experts. This conversation indicates that they are probably headed for trouble.

Supervisor: Jim, I think our company has reached a point where we need to consider new types of marketing. I'm thinking about some form of audiovisual production. We need to let people know about our new product line. That's why I want you to investigate our media options. Where do you think we should start?

Jim: Well, after we identify our audience and determine our goals and objectives, we need to come up with a script.

Supervisor: Great! It seems as if you have an understanding of this business, Jim. I'd like to see a script on a program for our new product line by next week. Think you can handle that?

Jim: Well, I'm not sure. Let me look into it.

Jim had better start his investigation immediately because chances are he doesn't know the first thing about how to prepare or evaluate a finished script.

THE SCRIPTWRITING PROCESS

Those who have little or no knowledge of audiovisual productions often take the scriptwriting process for granted. And because they have no background in audiovisual production, they fail to consider the importance of a script in relation to the final product. This chapter will provide an overview of the scriptwriting process, the various formats used to write a script and how to read them, accepted methods of payment for scripts and some suggestions on how to evaluate finished scripts.

Research

After a project analysis is completed and the audience is identified, the scriptwriter begins work. The first phase of scriptwriting is research, which takes a variety of forms including a review of print materials and current audiovisual programs about the topic. Writers also conduct interviews with the client, content experts and others who are knowledgeable about the subject. To enhance their understanding of the subject, it is also helpful for writers to observe the operations or services that will be covered in the script. Although many scriptwriters use only 10% to 20% of their research material in the final script, this extra information and effort are not wasted. By becoming expert on a topic, they are able to determine what information must appear in the script and what information is extraneous or unusable.

The Treatment

The next step is to write a treatment that describes in narrative form what the audience will see and hear in the finished production. During this critical phase, the writer develops a story concept and establishes the theme and visual style for the program. Many scriptwriters divide the treatment into major sections and estimate the amount of time that will be devoted to each of those sections. This gives the client a better feel for the length and flow of the program.

The treatment covers both visual and audio portions. Information concerning the visual aspects of the program would include major scene and set descriptions, special effects notations, camera movements (when appropriate), graphics elements, on-camera talent placement, and anything else that will enhance the client's understanding of what the program will look like.

The audio portion of the treatment covers key thoughts and phrases of dialog or voice-over narration. The writer also suggests the style(s) of music and where it should occur. If sound effects are to appear in the program, the scriptwriter notes what they are and when they will occur.

When preparing the treatment, writers use phrases like "The audience will see . . ." or "We will hear . . . ." This helps the client visualize program ideas and allows for an easy-to-read treatment.

Whether you write or approve a script treatment, be sure it is completely understandable. Identify the major ideas and concepts in the script and see how they interplay with the visual and audio aspects of the program. If any part of the script treatment appears confusing or inaccurate, correct it immediately. If an entire treatment concept is rejected or much of the treatment is altered, insist on a new treatment before continuing to the rough script stage. Do not allow a conceptual problem to slip by during the treatment stage. The problem will reappear in each subsequent stage of the script and will only be magnified in the final production.

The Rough Draft

Once the treatment is approved, the writer prepares a rough draft. This version of the script provides a detailed explanation of the program's visual and audio content. All generalizations contained in the treatment are made specific in the rough script; each shot is explained, narration is precisely written, scene locations are finalized, talent is characterized, and music and sound effects are detailed.

Scriptwriters have a choice of formats when writing the rough draft. For business and industry, most writers select one of the following: the split page, the storyboard or the corporate teleplay. Each format has its advantages and disadvantages; we'll review all three.

The Split Page

The split page (see Figure 3.1) is most often used for slide/cassette programs and speaker support presentations. It also remains popular with newscasters and advertising houses. For film and video, however, the split page presents one major disadvantage. As Donna Matrazzo states in *The Corporate Scriptwriting Book*, "By its very style, it encourages the writer to script the spoken words first, which is exactly the opposite of what the writer should be doing." Instead, the writer should write the visuals and audio simultaneously. Another problem is that the split page format encourages the client to read the narration on the right side and neglect the visual descriptions that appear on the left. We have heard more than one scriptwriter say

Figure 3.1: The split page script

53. MS of security fence.	ANNCR: Warnings and a variety of security devices protect against unauthorized entry.
	MUSIC continues under.
54. MCU of person inspecting air scrubber equipment.	ANNCR: Inspections of all types must be conducted on a routine basis.
	MUSIC continues under.
55. MS of person inspecting tank.	ANNCR: The site must be inspected for malfunctions, operating errors and discharges.
	MUSIC continues under.
56. MS of person monitoring well.	ANNCR: Monitoring and the results of this activity must also be reported.
	MUSIC continues under.
57. MS of employee being trained in use of fire extinguisher.	ANNCR: RCRA also calls for training of all employees in the handling of emergency equipment.
	MUSIC continues under.
58. 2-shot of employees checking lab samples	ANNCR: In addition, special precautions to prevent accidental reactions must be in place.
	MUSIC continues under.
59. MS of employees in fire truck.	ANNCR: In the event of an emergency, personnel must know how to respond...
	MUSIC continues under.
60. MS of employee with spill control equipment.	ANNCR: ... and know what equipment to use.
	MUSIC continues under.

that after the clients read split page scripts, they think that the words are good, but wonder where the pictures are.

The Storyboard

The storyboard script format (see Figure 3.2) is popular for film, multi-image and video productions. It is valuable because it allows the client to see what the visuals will look like. Individual drawings of each shot or scene are used instead of written descriptions. Appropriate audio information appears below each of the drawings.

This format is principally used by advertising firms for broadcast television commercials. Directors also use this format to visualize and plan complex segments of programs. It provides a step-by-step breakdown of the action that will occur in a scene.

There are two major disadvantages to the storyboard: it is expensive and time-consuming to produce. If a professional artist is not hired to draw the visuals, this format looks amateurish. Any alterations to the script require that the artist redraw some or all of the visuals. You can imagine how this adds to both cost and production time.

The Corporate Teleplay

The corporate teleplay (see Figure 3.3), with its various adaptations, is rapidly becoming the standard script format for video productions in business and industry. It provides the client and producer with scene numbers, exact locations and precise visual and audio descriptions. The visual description runs from the left to right margin. Below it, centered on the page, are the words that the on-camera talent delivers. In addition, both music and sound effects are centered on the page. Accompanying off-camera narration appears on the right third of the page. Specific transitions (i.e., cuts, dissolves, fades, special effects) appear before each scene. The primary advantage of the corporate teleplay is that clients tend to read the entire script because the layout is conducive to understanding the relationship between visuals and audio. Many find it to be a much more readable script format.

SCRIPT ELEMENTS

Narration communicates the content of the program and is only heard once, so for maximum impact, write in a conversational tone and use the active

28   MEDIA FOR BUSINESS

Figure 3.2: A storyboard

PROJECT: ___Technical Communications at Metropolitan State College___
DATE: _4-1-91_   PAGE _1_ OF _32_

VIDEO Fade to EXT. Long shot of West Classroom
AUDIO NATURAL SOUNDS

Cut to INT. Medium close-up of Tech Comm office door.
MUSIC: Fades up.

VIDEO Cut to INT. Close-up of window and Department name.
AUDIO MUSIC: Continues up.

Dissolve to GRAPHIC.
MUSIC: Continues under.
V/O NARRATOR: TECHNICAL COMMUNICATIONS. YOUR DOOR TO THE FUTURE.

Figure 3.3: The corporate teleplay

Cut to:

18. CU of conditioner or shampoo being poured into palm of hand.
        MUSIC: Continues up.
        NATURAL SOUNDS

Cut to:

19. MCU of stylist spraying hair. We see the mist lit in the background.
        MUSIC: Continues up.
        SOUND EFFECT: Spraying sound.

Cut to:

20. ECU of lips with ruby red lipstick.
        MUSIC: Continues under.

Cut to:

21. ECU of scissors cutting hair. Same shot as scene #1, but in focus. Pull back to LS and have narrator walk into shot. Frame him up in front and to right of chair with client. At end of On-camera, he turns toward client and stylist.
        MUSIC: Fades under.

        On-camera narrator:
        We all know how competitive the salon industry is today. The client base is shrinking. It's more expensive to develop new clients.

Cut to:

22. MCU of stylist and customer.
        MUSIC: Continues under.   ANNCR V/O:
        REPEAT BUSINESS IS
        HARDER TO COME BY.

Cut to:

23. Zoom in to CU of beauty products.
        MUSIC: Continues under.   ANNCR V/O:
        AND THERE ARE HUNDREDS
        OF PRODUCTS. . . .

voice. In other words, write the way people speak. Also, keep sentences short. Matrazzo suggests none longer than 15 words. This will prevent the pace of the show from becoming too slow or tedious.

Also, watch for tongue twisters and unfamiliar words. This simplifies the narrator's job and leads to greater audience comprehension. Always remember who your audience is. Avoid using technical jargon the audience does not understand. And be careful with slang; it is inappropriate for some audiences and quickly becomes dated. For instance, you would not write the following sentences for an older audience: "It was a totally tubular experience. Really off the wall, man." Not only does it contain words that are inappropriate for the audience, but the slang dates the program.

Some of the same common sense ideas hold true for music. In most cases, "easy listening" music in a program for teenagers would be inappropriate. Conversely, you would not use "heavy metal" music in a program for financial institutions.

Sound effects are another audio alternative. When used in an appropriate manner, they add realism and provide emphasis. The added sound effect of a telephone ringing, a gun firing or tires squealing may make the audience more attentive. However, be sure sound effects are true to the visuals. For instance, you wouldn't want to hear tires squealing if the car is on a dirt road.

As a client, you need to read and assess the rough script with an open yet critical mind. Remember, your goal is to translate this conglomeration of words and numbers so that you understand precisely what the final program will look and sound like. Most scripts require more than one rough draft and some go through as many as five revisions. Be prepared to pay additional costs for multiple script rewrites.

Before reading the rough draft or final script, remember that the script is written to be seen and heard, not read. This means that writers must think visually and write for the ear. In his book, *Images, Images, Images,* Michael Kenny says writers must write in sequences. Think about the last movie you saw. Can you recite any of the dialog? Probably not. But many of the visual sequences remain vivid.

The last step after the rough draft has been approved is completion of the final script. Once the changes made in the rough-draft stage are incorporated, the writer distributes the final draft to the client, content experts and any other people who must approve and sign off on the final script. Depending on the number of people involved, this can be a relatively short and simple process, or a long and complex one.

SCRIPT EVALUATION

When evaluating a final script, look for a theme, something that ties the program together and gives the audience something to remember. Unlike other reading matter, a script should be judged by how it sounds. Don't expect to read a script for a 10-minute program in half an hour. Study it. Make sure you understand it. Note any technical errors or unclear passages.

Remember that the script will read like no other written communication that you have previously read. Don't forget that incomplete sentences, slang, grammatical errors, oversimplified concepts and seemingly short sentences will be common. They are not mistakes, but techniques the scriptwriter uses to convey the information in a program.

Since scripts are also meant to be seen, make sure the visuals communicate the message. Don't rely on words; they will be forgotten long before the visuals. Don't overload the script with detail. No audiovisual program can adequately explain complex topics. That's a job better left to support people (content experts, trainers, etc.) and supplemental material (manuals, pamphlets, brochures, etc.).

If you are writing or reviewing a training script, be sure the program content addresses the measurable goals that are listed in the project analysis stage. (See Chapter 1, Effective Presentation Planning. For information on evaluating finished programs, see Chapter 14, Evaluation.)

SCRIPT PAYMENT METHODS

There are a number of different methods writers use to charge for a script. Matrazzo, in *The Corporate Scriptwriting Book,* lists the following: 1) the price-per-finished-minute charge; 2) the percent-of-budget offer; 3) the up-front bid; 4) the pay-in-three-parts contract; 5) the post-charge; 6) the what-the-market-will-bear charge; and 7) the flat-fee-plus-residuals charge. One other method is the hourly rate. This and the price-per-finished-minute charge are the most common. Hourly rates vary, based on the geographic location and demand, but clients should expect to pay between $25 and $60 an hour for the services of a professional, established scriptwriter. The price-per-finished-minute charge also varies based upon location and competition. (The current rate in Denver, for instance, averages about $160 per finished minute.)

Some scriptwriters prefer the pay-in-three-parts contract. This method of charging for scripts is becoming more popular because it provides the writer with cash in advance and thus facilitates research. It also ensures that the writer will receive payment for a treatment even if the client cancels the program. The other two payments provide for work done on the rough script and final draft.

A signed contract or agreement usually precedes any formal scripting work, although some scriptwriters still work under oral agreements. Any contract should precisely state the terms of agreement—due dates, methods, amounts and terms of payment; provisions for cancellation of services; proprietary and confidentiality concerns; rights to the finished material; extraneous charges; use of the script or program in a demo reel and other items. An equitable contract protects both the writer and the client.

SUMMARY

A good script serves as the inspiration for the producer and director and allows the client to visualize what a finished program will look and sound like. A poorly written script can cause cost overruns, demoralize the production crew and is rarely produced on schedule. So don't underestimate the importance of the scriptwriter in a program's success.

And now let's move on to a discussion of the audio portion of your media production.

4
Audio

Many people underestimate the importance of the audio portion of a media production, as the following all-too-realistic dialog suggests:

Client: You've done a great job with the visuals for this program. Now all we have to do is add some music. Use anything you want. I don't really care what it sounds like.
Producer: I've got a good idea about what to use. And the musical selections I've chosen should only cost about $200.
Client: Really? That's awfully high. Why don't you just use something off this album I bought the other day? It has some great songs.
Producer: I'm sure it does, but the music needs to mesh with the visuals. And besides, you don't want to get sued, do you?

The client here is forgetting how audio complements a program's images by clarifying what the visuals cannot communicate. In this chapter, designed to convince all such doubting Thomases of the intrinsic importance of sound to a media production, we will outline the basic audio elements. We will also include a guide to working with them and will review some common audio equipment.

THE ELEMENTS OF AUDIO

Audio has five basic elements which producers use alone or in conjunction with one another.

The five basic elements are the spoken word, music, sound effects, natural sound and silence. It is important to know how these elements are used, their respective advantages and disadvantages, and associated costs for each. We'll look at them one by one.

Voice

The spoken word—or human voice—is the most common type of audio. Voice can be recorded as voice-over or by on-camera actors. Voice-over narration is audio that is recorded and then synchronized with visuals. The narrator is not shown. (More than one narrator may be used in a program. In fact, use of two narrators in a long program helps break it up. Just be sure to use two distinct voices or employ both a man's and a woman's voice.) If you use on-camera talent, both image and words are simultaneously recorded on film or videotape.

How do you go about contacting and choosing narrators? First, you must decide whether to use professional or amateur talent. Amateurs are appropriate when you cannot afford professionals or when you are aiming for a non-actor's authenticity. Use amateur talent when that person lends credibility to the script. Since amateurs do take much longer to record and since the adage "time is money" holds especially true in audio recording, we advise hiring professional talent.

Next you have to decide whether to hire union or non-union talent. Although a union card does not necessarily ensure a quality narration, it does indicate that the narrator considers himself or herself a serious professional. Be aware that some audio studios have strong preferences in this regard. This varies across the country, but most major metropolitan centers (especially those east of the Mississippi) are liable to have strong unions. Mixing union and non-union talent can also be a tricky proposition. It usually works once; thereafter, non-union people often join the union.

Both on-camera and voice-over narrators can be contacted in different ways. Audio studios can provide you with names of people they have used in past productions. The Screen Actors Guild (SAG) and American Federation of Radio and Television Artists (AFTRA) can offer advice and give you names of their members. Most major cities have a listing for SAG or AFTRA in the phone book.

Talent agencies are another option. They represent actors and actresses and actively seek work for these professionals when a producer needs talent. They suggest people, set up auditions and provide you with narrator demo tapes. These tapes provide examples of the audio work the narrator has previously done. A good demo tape will also include different voice characterizations, celebrity voice impersonations, accents and dialects the narrator can do.

Another source of voice talent is casting agencies, which are similar to talent agencies, except that by contacting all talent agencies, they provide their clients with a larger and more diverse talent pool. (Again, most talent and casting agencies are located in major metropolitan areas. Consult the phone book for their numbers).

One final source for narrators is radio stations. Radio people who do voice-over narration are usually affiliated with a talent agency or audio studio. Some make excellent narrators; others don't. We advise you to audition radio talent just as you would actors, and make sure the person is capable of reading copy that lasts more than a few minutes. Also, remember that disc jockeys have recognizable voices. This could be detrimental to your program because the audience might be distracted by associating a disc jockey's voice with a station, a type of music or a product, instead of listening to your program.

The best way to choose a narrator is to see and hear that person at work. If possible, have the prospective narrator read and record a section of the script you have chosen. This is sometimes more helpful than listening to a demo tape or seeing a person's demo reel. Some people are excellent voice-over announcers, but do not look good on camera. If you require an on-camera narrator, record an audition and judge whether that person comes across well on videotape or film. (Record the audition with a camcorder. This eliminates the need to spend a lot of money.)

Costs for narrators vary. For a five-minute program, fees for a professional narrator might range from $100 to $1000. Factors affecting cost include use of union or non-union talent, length of program, medium in which the program will be produced and the audience (in-house, broadcast, non-profit, educational, etc.). Celebrity narrators normally charge more for their services because of voice recognition.

With respect to script format, most narrators prefer to work with a script that is double-spaced and typed on the right half of the page with line numbers placed on the left half of the page (see Figure 4.1).

Remember that it is almost impossible to record a quality narration in somebody's office. You need to record your narration with professional equipment and in a quiet environment, whether you are using in-house facilities or a commercial audio studio.

Music

Music sets the mood, evokes emotion or provides a comfortable background for the rest of the audio, preferably without distracting the viewer from the content of the program. Music can also signal the end or beginning of a program or section and it can indicate a break in content. There are three main

Figure 4.1: A narrator's script

COUNTRY SCHOOL V/O SCRIPT

<u>Open: 45-50 seconds</u>

1	Heritage. The word evokes memories
2	of the past—of simpler, less
3	complicated times. Of times we
4	cherish. Sometimes we recognize our
5	heritage in the form of a building
6	or object. Other times a sight, a
7	smell or a texture reminds us of our
8	past. And whether our heritage was
9	born in the old country, or nurtured
10	in the new world, it serves as a link
11	between generations. Today, in
12	Lakewood, we have the opportunity to
13	preserve part of ourheritage for
14	ourselves and future generations.

<u>History: one minute</u>

15	One room school houses, like the
16	Country School, played a major
17	rolein our state's history. The
18	values and traditions taught in these
19	buildings became part of our
20	heritage. (Pause 2-3 beats).

sources for the musical soundtrack of your media program: original music, existing copyrighted music, and selections from music libraries.

Original Music

The great advantage of original music is that it is composed to match your spoken words and visuals. It emphasizes words, sentences and ideas, or highlights pauses in the narrative. This customized quality is more difficult to achieve with library music or existing commercial compositions.

The first thing to do if you decide to commission original music is shop around. Ask to hear compositions; get cost estimates. Choose a person with whom you can work closely, give the composer a copy of the script as far in advance as possible and describe the type of music you want in the program. If you have a musical background, suggest the types of instruments to be used. Let the composer know what style you envision—symphonic, rock, jazz, futuristic, big band, or whatever. If your knowledge of music is limited, try to describe the soundtrack in terms of an existing musical piece (a Beach Boys song, a Chopin prelude, or the theme for "Hill Street Blues").

Another way to describe soundtracks is in terms of emotions or moods. Most composers realize that a sad or foreboding mood calls for a minor key. By being as clear as possible about what you want, you can greatly simplify the composing process. If, however, you have implicit trust in your composer, let that person interpret your script and compose accordingly.

One main reason not to commission music is cost. Expect to pay from $500 to $3000 per minute for original compositions. (Be sure to figure in the cost of hiring professional musicians to record the soundtrack.) Second, unless otherwise agreed to in a "work-for-hire" contract, the composer retains the copyright and use of the music. This means the music may be used for other purposes at the composer's discretion. Third, what happens if you dislike the musical score? You still have to pay the composer, and either you start over or make revisions. Either way, you'll probably end up paying more than you expected.

Using Copyrighted Music

Recording copyrighted music is another option when you are putting together your soundtrack. Unfortunately, this has a number of disadvantages. First, you need to obtain permission from the composer, record company, publishing house and any other people or entities who hold rights to the music or performance before you can use the piece. In addition, fees for use of commercial music can be prohibitive if you're working with a limited budget. Finally, audiences associate specific emotions and situations with well-known songs.

Take the Beatles' song "Eleanor Rigby," for example. Many audiences associate this song with the movie "Yellow Submarine." Some people might recall where they were or what they were doing the first time they heard it. For others, it might evoke emotions unrelated to your program. Because a song can distract, rather than focus attention, its inclusion can be a detriment to your program.

This does not mean that copyrighted, commercial music is taboo. If a specific song correlates with the visuals and adds meaning to the program, try to obtain permission to use it. This means that you must get synchronization rights (permission to use a particular composition) or mechanical rights (permission to use a particular recording). The easiest way to obtain these clearances is to hire a permission rights service. You will pay an average hourly fee of $50 for research, negotiation and other services. A representative from the permission rights company will contact the holder(s) of the copyright and publishing rights. They will negotiate a price for use of the piece in your program which you may then accept or reject.

But remember, even if you are willing to pay exorbitant prices, the owners of the copyright are not obligated to grant you permission to use the piece. Also, you may only have permission to use the piece for a limited time period. Be clear on the use of copyrighted music.

Music Libraries

Commercial music libraries are the most popular source of music for media programs. These are collections of original, prerecorded musical pieces that companies pay composers to write. The company that owns the rights categorizes, markets and sells the music to media producers for use in their programs. There are hundreds of music libraries produced in the U.S., Canada and Europe, and the choices vary. Virtually any type of music, from rock, jazz and pastoral to industrial, classical and ragtime, is available through music libraries. And you may choose the audio format of your liking—record, reel-to-reel tape, cassette or compact disc. Most commercial libraries provide sample tapes of their music. Many audio studios also offer a variety of libraries that you may sample for use in your program.

There are two ways to purchase music from libraries: "needle drop" and leasing. For a fixed fee, you may choose selections from a variety of libraries. This fee is called a "needle drop" rate. Although this term is associated with tape and compact discs, it specifically refers to placement of the turntable stylus on a selection of music from a record album. Prices range from $30 to $150 per needle drop for most industrial, nonbroadcast productions.

The other method of purchasing music is to lease the library on a yearly basis. Fees vary from $500 to $2000. This allows the lessee to use any selection of music in the library an unlimited number of times for media productions. Variations in leasing agreements do exist, but they are spelled out in the contract. Instead of leasing their music, some companies offer a "buyout" of their libraries. For a one-time, fixed price, this agreement allows the buyer to use any selection in the music library an unlimited number of times forever. Buyout libraries usually contain fewer selections than those that are leased. And, in many cases, the music quality is inferior to that of the larger, more comprehensive libraries.

The principal advantage to working with music libraries is that their music is less expensive than original music or popular copyrighted music. They also provide a great variety of musical selections in different styles, versions and time lengths. Many music libraries also feature variations on themes with different types of instrumentation. These variations can add to the continuity of a program.

The downside to using music libraries is that anybody who can afford the price of a particular selection can use it. In addition, radio and television programs sometimes feature popular selections from music libraries. This can make your soundtrack sound too familiar. However, even the most popular music library selections tend to make media programs sound more professional.

Other Audio Elements

Sound effects are another audio element that enhances media production by adding realism and providing emphasis. They may be obtained in a number of different ways. If time is not a factor, sound effects can be recorded on a high-quality portable cassette or reel-to-reel recorder. They can also be recorded in a studio by individuals. We knew a person who could vocally mimic almost 30 different animals. She could also imitate creaky doors, race car engines, jet noises and a host of other sounds. She provided us with a small sound effects library at virtually no cost.

40   MEDIA FOR BUSINESS

Figure 4.2: An E-MAX sampling device

Courtesy of Steve Allen.

The advent of new recording technologies has introduced a simpler method to produce and record sound effects. Through the use of sampling devices (see Figure 4.2), practically any sound effect is now available to the producer. Mark Petersen of Loch Ness Productions in Denver describes sampling devices. "Samplers are keyboards which make and play back digital recordings. They can be used for sound effects and to reproduce other musical instrument sounds. For instance, you can record an animal sound on location, then record it onto the floppy disc of a sampler. You can do the same with musical instruments. Or, you can record sound effects from other sources—records, tapes, compact discs—onto the sampler's floppy disc. Samplers are versatile because the keys allow you to actually play the sounds. You can also change pitch, frequency and tone, or repeat the sound a limitless number of times and manipulate it in many other ways."

The most common way of obtaining sound effects is to purchase them from a commercial library, just as you would buy stock music. They cost as little as $7 per effect—much less than selections from music libraries. There are a number of existing sound effects libraries from which you may obtain sample tapes.

Natural sound, also referred to as ambiance, is another audio element. Natural sound is the audio recorded while shooting videotape or motion picture film. It is the sound—either man-made or environmental—that occurs naturally at the time of recording. If you videotape an interview in a park by a busy thoroughfare, in addition to the interview, you may hear traffic sounds, birds chirping, children shouting and other extraneous noises. All of this is natural sound.

Silence is the forgotten audio element. When used right, it can be an effective attention-getting device. If overused or used in an improper manner, however, audiences assume silence is a mistake. Use it judiciously.

AUDIO EQUIPMENT

Audio equipment comes in various forms, but it has two basic functions: it records and plays back sound. But before we begin to review equipment, let's note a few basics, such as the replacement of analog by digital recording. Until a few years ago, analog recording methods were the norm. However, this type of recording is noisy in comparison to digital—unwanted characteristics like tape hiss and distortion are more evident. Digital audio recording has been able to eliminate many of the problems associated with analog. In simple terms, digital recording stores the original analog signals in a binary form. The electrical signals are encoded on audiotape and then decoded during playback. When played back, digital recordings consistently produce the same quality of sound. They do not deteriorate with repeated playback, as analog recordings do. Digital's one main disadvantage, however, is cost. In terms of tape stock, equipment and editing time, it exceeds the cost of analog recording.

Also important to our discussion is an understanding of basic recording principle. Audiotape is the medium for sound recording. It is manufactured in various sizes and thicknesses, depending on recorder type. Although new recorders no longer make this a universal rule, generally, the larger and thicker the tape, the better the recording quality. During the recording process, tape passes across the recorder heads at specific speeds. These speeds are expressed in inches per second (ips). Record decks usually have two or three speeds at which they record. Recording speeds start at 15/16 ips then double in speed at each interval. The range of recording speeds is: 15/16, 1 7/8, 3 3/4, 7 1/2, 15, and 30 inches per second. In most cases, the faster the recording speed, the better the audio quality.

Recorders

During the recording process, music, words or sound effects are first recorded onto audiotape, or in the case of video, onto the audio portion of the videotape. When using professional audio reel-to-reel recorders (see Figure 4.3), 1/4-inch, 1/2-inch, 1-inch or 2-inch tape is used. The recorders transport tape through the machine at speeds of 7 1/2 ips or faster. This recording equipment produces the best quality sound and allows you to put audio information on multiple tracks of the tape. For instance, a four-track recorder (see Figure 4.3) can store four pieces of sound on one tape. Reel-to-reel recorders with more tracks have even greater recording capability. Not surprisingly, the more sophisticated the equipment, the greater the cost of recording.

Audio is produced on cassette recorders when reel-to-reel decks are not available or when cost is a prime consideration. Consumer cassette decks use 1/8-inch tape to record two tracks of audio at 1 7/8 ips. Some industrial and professional cassette recorders have three or four tracks and an additional recording speed of 3 3/4 ips.

Figure 4.3: 4-track reel-to-reel tape deck

Courtesy of TASCAM, TEAC Professional Division.

Playback Equipment

Playback equipment includes turntables, cassette decks, reel-to-reel players and compact disc players. Whereas many cassette and reel-to-reel decks have both recording and playback capability, turntables and compact disc players can only play recordings.

There are two types of turntables: belt driven and direct drive. Direct drive turntables are more consistent because they do not rely on belts (which can wear out) to move the turntable platter. All turntables have a tone arm that consists of a stylus and cartridge. The stylus, also called a needle, tracks across the grooves of an album, picking up vibrations. It transfers this information to the cartridge which converts the information to sound. Since the stylus must come in contact with the record each time in order to produce sound, both the record and stylus will eventually wear out. While reel-to-reel and cassette players deliver high-quality audio without using a needle, they also subject the recording to wear since the playback heads must touch the tape in order to produce sound.

Compact discs offer an alternative to turntables and albums, cassettes and reel-to-reel recordings. A laser beam reads the surface of the disc and converts the information to sound. No physical contact occurs; therefore, discs maintain their fidelity for a longer period of time. In addition, it is easier to fast forward and reverse compact discs. They also provide easier access to specific selections.

Microphones

Microphones, which are used to record voices or sound effects, can be categorized according to directionality—the pattern by which they pick up sound (see Figure 4.4). Omni-directional microphones pick up sound from all directions. Bi-directional mikes detect sound from the front and back, but not the sides. Uni-directional mikes pick up sound from one direction. Cardioid mikes have a heart-shaped pickup pattern.

Figure 4.5 shows some examples of the sizes and types of microphones available. The most common are lavalier, cardioid and shotgun microphones. The lavalier, also called a tie clip mike, is typically used in corporate news situations. The cardioid is hand-held, mounted on a stand or used on a boom to pick up vocals. A shotgun mike has a highly directional pickup pattern. The

Figure 4.4: Basic microphone patterns

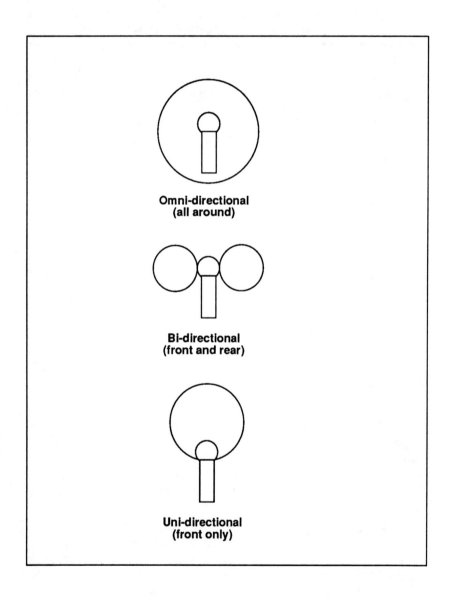

Figure 4.5: Examples of microphones (left to right: hand-held cardioid, shotgun with zeppelin, lavalier, cardiod)

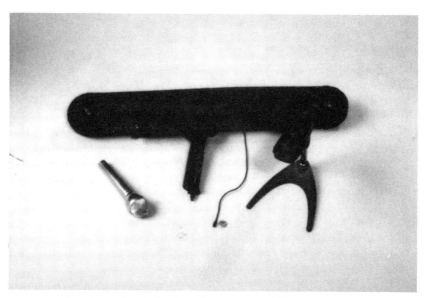

operator aims it at the desired audio source as if it were a shotgun. Most microphones have wires which connect them to a recording source. However, wireless mikes are becoming more popular, reliable and affordable, and they are simple to operate. The microphone transmits a signal on a specific frequency to a receiver, which accepts the signal and relays it to the audio input of a recording device.

There are, of course, countless other types of microphones and peripheral equipment—wind screens, booms, stands and other devices. Always use the right type of microphone for the situation, and place the microphone as close to the audio source as possible. Another alternative is to hire a professional sound person. Stanley Alten's book, *Audio in Media,* is an excellent source for more specific information concerning microphones and their use.

Other Equipment

Parametric equalizers, phlangers, phase shifters, noise gates, delay units, enhancers and a range of other hardware can be found in a well-equipped studio. But the most important piece, in addition to recording devices, is the mixer (see Figure 4.6).

Figure 4.6: A 6 x 2 audio mixer

Courtesy of TASCAM, TEAC Professional Division.

Mixers are invaluable in the recording studio and in the field. They combine audio from different sources—turntables, cassettes, compact discs, microphones, etc.—and send those sounds to a recording device. Mixers are defined by the number of lines they can take in and the number of lines they can output. For instance, a 6 x 4 mixer can receive information from six distinct sources and output that information to four lines (to a four-track recorder, for example). Sophisticated mixers, those used to record commercial music albums, can handle up to 64 different audio sources. When used in conjunction with multi-track recorders, the options they offer are limitless.

AUDIO STUDIOS

When in-house facilities do not exist or are inadequate, the producer must turn to a commercial audio studio to make a soundtrack. Commercial studios can be found in all metropolitan areas. They offer facilities for recording voice-over tracks, music, and sound effects. Professional studios have soundproof rooms and furnish high-quality microphones for recording narration. Some have specially designed rooms for recording musicians. And each studio has a professional mixer as part of its standard equipment.

Studios offer high-quality equipment and expertise that few people or companies possess. Since they are in competition with one another, they tend to satisfy their customers by producing quality soundtracks within budget and on time. But they can be expensive. Most studios charge from $50 to $150 per hour for studio time. This does not include material costs or other fees like music library prices, royalties, musicians' salaries and narrator fees. As you should for any other service you purchase, shop around. Compare rates and facilities. If you find someone with whom you like to work, consider that in your decision. Other studios may be cheaper, but their personnel may not be as technically adept or creative.

Finally, when estimating the amount of studio time you will need, determine the type and number of sound elements you need. If you expect it to be complicated (two voices, two or more musical pieces, sound effects), figure one hour in the studio for every minute of your program. That may sound inflated, but it is a good way to estimate studio costs.

STAND-ALONE AUDIO PROGRAMS

As we have previously discussed, audio most often accompanies visuals in media programs whether for a slide/cassette presentation, multi-image show, videotape or film. But many companies use stand-alone audio programs for educational, informational and motivational purposes. Salespeople can review sales techniques by playing an audio program on their car cassette players while driving to an appointment. Audio recordings of lectures are far more effective and cost-efficient than videotaping a speaker. And many motivational programs on topics such as weight loss and building self-esteem are best presented in the form of an audiotape.

SUMMARY

Audio is an important element of many media programs. It clarifies content, enhances realism, serves as a bridge between segments and increases emotional impact. Unfortunately, it is too often ignored. The good producer understands the value of audio, makes sure it is properly recorded and uses it for maximum effectiveness.

In our next chapter we will take a look at graphics in media.

5
Using Graphics

Stop for a minute and ask yourself how many graphics you've seen today. Think about it. In the newspapers, on television, in magazines, just about anywhere you look, graphics are being used to communicate a message. This is no accident. Graphics work, and the reason they work is because people can relate better to what they see than to what they hear or read. This holds true for the business presentation just as it does for the evening newscast or a magazine article. Used skillfully, graphics can simplify a complex process, illustrate an idea, drive home a point, or provide a much needed touch of humor. Business graphics can make the difference between confusing the issue and making your point. They allow your audience to "see" the facts instead of just hearing them.

Before we talk about the types of graphics used in business presentations, we should define the term. The Random House College Dictionary defines the word "graphic" as: "Giving a clear and effective picture; vivid." The business graphic should do just this, give a clear and effective picture of your message, whatever that message may be. Graphics are communications tools just like the presentations they support. They are primarily used to communicate facts, figures, concepts and ideas that would be difficult to communicate by verbal or written means. They are a powerful tool for the business communicator.

TYPES OF BUSINESS GRAPHICS

Graphics are called on to serve many purposes in the business presentation and many different types of them are commonly used by business communicators. These include charts and graphs, illustrations, animation and cartoons. Let's look at each of these in more detail.

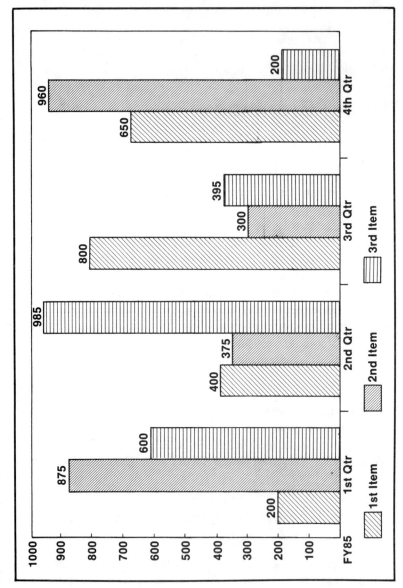

Figure 5.1: Bar chart, example of X-Y axis chart

Charts and Graphs

The most common graphics in business presentations, charts and graphs are used to illustrate the relationship between two or more related variables. In business presentations, variables can represent anything from people to sales figures to production quotas.

There are literally hundreds of different types of charts and graphs used for business presentations. In fact, nearly every business develops subtle variations that are unique to its organization's needs. Most charts and graphs, however, fall into the three basic categories discussed below.

Charts and Graphs Using an X-Y Axis

The familiar bar graphs, line graphs and area charts that complement business presentations everywhere are all based on an X-Y axis. (See Figure 5.1.) Charts such as these are ideal for illustrating a series of quantities recorded

Figure 5.2: Example of a pie chart

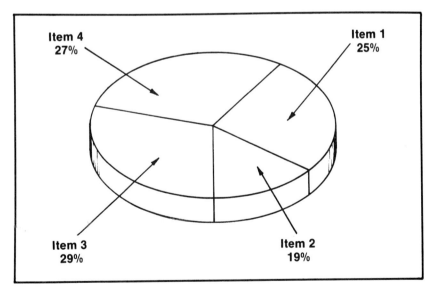

over time. An example of this would be a chart illustrating a company's sales figures for each month of the fiscal year. X-Y axis charts are also useful for comparing two related factors, such as costs for an automated manufacturing process versus costs for a manual manufacturing process.

Pie Charts

Pie charts (see Figure 5.2) have remained relatively unchanged since they were first used in the late 1800s. The classic pie chart is a circle with individual segments identified and quantified as a percentage or fraction of the whole. While the word pie chart suggests a circular design, pie charts can also be constructed from other shapes. Dollar bills, countries, states, and even pies can be used, depending on the subject matter involved.

Pie charts should be used to show individual sections which, when combined, create a whole. Pie charts can be used to depict a budget breakdown, a sales profile, audience composition, or anything that can be represented as a sum of its component parts.

Figure 5.3: Example of a table

1989 YEAR TO DATE OVERVIEW			
MONTH	YTD EXPENSES	YTD BUDGET	YTD VARIANCE
JANUARY	$18,129.52	$11,004.00	($7,125.52)
FEBRUARY	$78,024.81	$25,257.60	($52,767.21)
MARCH	$79,293.59	$28,925.60	($50,367.99)
APRIL	$90,121.47	$39,091.20	($51,030.27)
MAY	$99,351.28	$43,491.20	($55,860.08)
JUNE	$117,535.10	$53,491.20	($64,043.90)
JULY	$217,105.65	$62,491.20	($154,614.45)
AUGUST	$217,105.65	$107,491.20	($109,614.45)
SEPTEMBER	$217,105.65	$117,491.20	($99,614.45)
OCTOBER	$217,105.65	$129,591.20	($87,514.45)
NOVEMBER	$217,105.65	$179,091.20	($38,014.45)
DECEMBER	$217,105.65	$189,091.20	($28,014.45)
TOTAL	**$217,105.65**	**$189,091.20**	**($28,014.45)**

Tables

Tables (see Figure 5.3) is a sort of catch-all category for business graphics. As defined in Nigel Holmes' *Designers Guide to Creating Charts and Diagrams*, a table is "An arrangement of information, often numbers, into columns or other organized ranks or groups. . . . " Many different types of tables are used in business communications. Examples include organizational tables that illustrate the hierarchy of a company's top management and flow tables that illustrate the flow of a process or system. Tables are also used to organize a body of raw data, such as test results, into an understandable format.

Applying a little creativity to the design of charts and graphs can make them even better communications tools. For instance, a chart illustrating sales figures with a simple bar can be improved if the bars are depicted as stacks of coins (see Figure 5.1). This makes the graphic even more relevant to the subject matter and helps to increase its effectiveness.

Illustrations

An illustration is a drawing or representation of something—a building, a person, a machine, a component. This type of graphic can be very effective in meeting a variety of business communications needs.

Illustrations are ideal for depicting items that cannot be easily photographed. These can include exploded and cutaway views of parts, machinery, buildings and equipment; artists' renderings of unconstructed buildings, facilities and products; and drawings portraying functions and operations that are not visible to the human eye (such as osmosis or the firing of a spark plug within an automobile engine).

Illustrations can also be used to enhance a program by interpreting the subject matter in an artistic manner (see Figure 5.4). Sketches of people, places and things are often more interesting and effective than photographs in certain media presentations. Illustrations of this kind are often combined with text to help the audience "see" a thought or concept.

Cartoons

Cartoons provide a good way to inject a little humor into an otherwise dry business presentation. They can also help the audience focus on a topic or issue that you would like them to remember long after the presentation is over.

54 MEDIA FOR BUSINESS

Figure 5.4: Example of an illustration

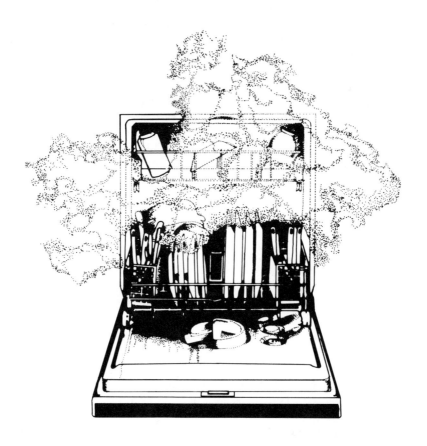

Also, people love to laugh, and making your point with an amusing cartoon is a sure way to warm up your audience.

Two minor cautions are needed here. First, be careful about copyright violations when using cartoons from a newspaper or magazine. There are limits to "fair use" of copyrighted materials and cartoons are subject to all applicable copyright laws. Second, when creating original cartoons, remember that being funny isn't easy. Designing effective cartoons requires a great deal of skill and understanding. Something that seems genuinely funny back in the office may be received quite differently in the boardroom. When possible, call on the services of a professional cartoonist.

Animation

Animation is a graphic technique that is limited to the "motion media," which include film, videotape, and to a degree, multi-image presentations. The value of animation to the business communicator is its ability to portray movement on the screen. Like illustrations, animation is often used to show things that cannot be photographed, or to visually enhance a presentation.

Through animation, obscure principles and phenomena such as the inner workings of machinery and equipment, the flow of blood through the human circulatory system and the movement of sound waves through the air can be clearly demonstrated.

Animation may also be selected for its entertainment value. If one were trying to show an audience the proper way of performing a certain task, a videotape showing real people at the task might bore an audience to distraction. An animated sequence, however, might be amusing, captivating, and above all memorable. Animation can also be used to simplify complex processes or procedures. And, animated characters can perform actions and suffer consequences that would be dangerous for humans and other living things.

COMMON GRAPHICS PITFALLS

There is no doubt that the skillful use of graphics can improve virtually any business presentation. But when graphics are abused and misused they can actually make a program less effective by detracting from, rather than enhancing, the message. Following are some common pitfalls to avoid when planning your use of graphics.

Conveying Too Much Information

By definition, graphics should present a clear and effective picture of the message they are designed to communicate. All too often, business communicators feel they must cram as much information as possible into a single graphic. Inevitably, this results in a confusing visual that is neither clear nor effective. With graphics, more is definitely not better.

Avoid cluttering your graphics with information that does not support the message. Try to structure the graphic so that it paints a single, clear, concise picture. Eliminate superfluous, non-essential data that only cloud the issue. The finished graphic should require a minimum of explanation. The audience should be able to immediately recognize and interpret the information and draw appropriate conclusions.

Using Color Improperly

Color can be a valuable tool for enhancing business graphics but used improperly it can draw attention to itself and away from the message you're trying to communicate. A major problem with many business graphics is the over use of color. Many people seem to feel that if they use as many colors as possible, their presentations will be more effective. Nothing could be further from the truth. Strange as it may sound, the colors you use in your presentation should be invisible to the audience. This means that the audience should be concentrating on your message and not on the colors you selected. If a person comments on the colors you used, he or she wasn't concentrating on the content of your presentation.

A good strategy for incorporating color into your program is to select the minimum number of colors you need and stick with them. Maintain color consistency throughout; don't switch color schemes on every other graphic. Using the same colors throughout the presentation will give your program a professional, cohesive quality that the audience will appreciate.

Ignoring the "Safe Title" Area

Many business presentations produced on 16mm film and 35mm slides are eventually transferred to videotape. If the graphics produced for these programs were not designed to fit within the television "safe title" area, the graphics may appear cropped when viewed on a television screen. The "safe title" area refers to a section in the center of the 16mm or 35mm film frame that the graphic must fit into if it will ultimately be shown in video format. Anything falling outside of the safe area may be cropped off during the transfer. This is because the aspect ratio (height to width ratio) for a television image is different than the aspect ratio for either film format. If film-to-video transfers are common at your facility, it is good practice to design all presentation graphics so that they will fit into a TV safe title area. It is a relatively simple matter to design graphics so that they meet this requirement.

PRODUCING BUSINESS GRAPHICS

The best way to make sure that your graphics are as good as they can be is to have a graphic artist create them for you. Original illustrations, technical drawings and cartoons are most effective if they are produced by a professional. It is true that many business graphics are designed and produced by secretaries, engineers, planners and others with no formal art training whatsoever. But these efforts pale in comparison with a professionally designed and produced communications package. Most major corporations maintain in-house media arts departments and much of their time is spent producing graphics for business presentations. If these services are not available within your company, seek out a commercial art vendor that specializes in presentation graphics.

Desktop Publishing Systems

Not all presentations will merit the attention of media professionals. For the lay producer of slide and overhead transparency presentations, a desktop publishing system is a dream come true. Professional-looking graphics can be produced in minutes without ever having to leave the comfort of your desk.

Desktop publishing systems are actually nothing more than a standard personal computer equipped with an appropriate software program and output device (preferably a laser printer). More sophisticated systems will include a mouse or a writing tablet and stylus for creating freehand drawings.

The heart of any desktop publishing system is the software package. There are many excellent programs on the market that will more than satisfy the needs of most business communicators. These programs range in price from $100 to $500. The higher-priced programs will typically feature greater graphics capabilities. Consult your local computer software supply house for a demonstration. However, before you rush out and buy a desktop publishing system, we should explain just what it will and won't do.

First, a look at what it will do. A well-equipped desktop publishing system will produce professional, high-quality charts, graphs, and tables that are suitable for virtually any business presentation. Graphics produced with desktop publishing systems can be used for overhead transparencies, 35mm slides, and, in a pinch, film and video presentations. With the proper printer, either black-and-white or color graphics can be created. Most systems will even design the charts and graphs by themselves. All you need to do is type in the figures and tell the system what type of graph you want. Systems equipped with a mouse or stylus will allow you to create freehand drawings directly on the computer and print them with your output device.

Now here's the bad news. Even a desktop publishing system will not turn you into a graphic artist. The creation of illustrations, cartoons and many customized charts and graphs will still require the services of a talented graphics specialist. We know that a computer will only do what you tell it to. And if you're not a graphics artist, chances are you'll probably tell your computer to do the wrong thing, graphically speaking, at one time or another. Many corporations are now equipping their professional illustrators and graphic artists with desktop publishing systems so that they too can take advantage of the creative possibilities offered by these systems.

Computer Graphics

It may seem confusing, but computer graphics and desktop publishing are actually two different animals. Both are computer-aided methods of creating presentation graphics, but the similarity ends there. Desktop publishing systems are designed for use by nonspecialists and utilize personal computers to create their product. Computer graphics systems are designed for use by trained professionals and often require specialized hardware specifically designed for this purpose.

Computer graphics are utilized for a wide range of presentation visuals, including everything from simple text-only slides to fine art graphics. Computer graphics are nearly always created in color, and are of higher visual quality than graphics produced on desktop publishing systems. In the hands of a skilled operator/artist, some high-end computers can create images that approach photographic quality. Imagination and budget are the only limitations when employing personnel and equipment of this caliber.

Many larger corporations are so impressed with this technology that they have invested heavily in computer graphics equipment, training and staff for their in-house graphics operations.

Animation

In the case of animation and certain other types of graphics for film, video, and multi-image presentations, you will have no choice but to seek out professional help. Producing graphics for these media often requires specialized equipment, knowledge, and expertise. Either your in-house media productions department or an independent production company will have access to both the necessary equipment and expertise to handle most requests.

SUMMARY

There are many types of graphics available to today's business communicator. Charts and graphs, illustrations, animation and cartoons can all be used to enhance and facilitate the delivery of your message. The ultimate goal of any business graphic (or presentation for that matter) is to create a clear and effective image. Remember to keep this in mind when designing graphics for your presentation. Eliminate extraneous information that only serves to complicate and confuse the issue. Make an effort to design your graphics so that their message is clear to the audience with a minimum of explanation or interpretation.

Finally, give serious thought to enlisting the services of a skilled graphic artist for your more important presentations. Many graphics can be designed and produced by non-artists with the assistance of desktop publishing systems. But overall, these efforts will not measure up to a professionally conceived and produced presentation package.

Next, in our last chapter in this section on processes used to produce media, we'll consider print.

6
Print for Media Presentations

Supplementing a media program with printed material can increase the effectiveness of the program and help meet goals beyond the scope of a single presentation. This means printed materials can be a valuable tool for media professionals who are continually faced with complex communications challenges.

In this chapter we will both discuss the role of print in media programs and provide a series of concrete examples of how and when to use it. But first, a look at what corporations expect from media programs.

Most business-related media programs have multiple goals. It is not uncommon for a business to request a single media package to train, persuade, inform and change the behavior of their employees. To the business person, who may be unaware of the limitations of a media presentation, this may seem like a reasonable request. To the professional communicator, however, it presents a very real set of problems. Here's why.

Let's think of a corporate media presentation as a tool—not unlike a wrench. Each wrench is designed to fit a certain sized bolt or nut. If the wrong sized wrench is used to remove a bolt, either the bolt or the wrench itself may be damaged in the process. Like the wrench, a media package is a tool designed to perform a specific task—training, selling, persuading, etc. When a media tool is expected to perform more tasks than it is capable of, either the media package or the audience suffers.

Suppose a company is introducing a new product line and produces a videotape to introduce it to the sales staff at a sales meeting. There are two goals the company would like to achieve during the meeting: to get the sales staff motivated to sell the new product, and to provide the staff with detailed product specifications. In other words, the purpose of the meeting is to both motivate and inform the sales staff about the new product. This is a tall order for a single media program because a different treatment, or media tool, is required to accomplish each goal. A motivational program must be upbeat and appeal to the audience on an emotional level. An informational program, on the other hand, is typically slower paced and filled with technical information that would be out of place in a motivational treatment.

How do we satisfy both of the company's goals without compromising the media program or cheating the audience? This brings us to the subject of print, the focus of this chapter. One way to accomplish both objectives is to produce a motivational videotape for presentation at the sales meeting and to distribute a brochure to the sales staff that contains the product specifications.

TYPES OF PRINTED MATERIALS

There is a variety of printed materials suitable for use with media programs. These include single- or multiple-page handouts, pamphlets, briefing books and brochures. There's nothing revolutionary about any of these. The idea here is not to reinvent the wheel, but to combine two effective media into a single integrated package that is capable of accomplishing multiple communications goals. Let's look at each in more detail.

Handouts

These are the most basic of all possible printed materials. They are nothing more than one or several pages of information provided to the audience either before or after the media presentation. Handouts are excellent for disseminating background information or technical details that are beyond the scope of the media presentation.

Pamphlets

These are the next step up from handouts and come in a variety of formats. They are generally fancier than handouts and may include graphic elements such as drawings or photographs. They may also be of coated stock, contain multiple folds, or feature complex design elements.

Briefing Books

These are unique in that they provide, in printed form, the same material covered during the media presentation. Briefing books are most often used to support slide and overhead programs.

Brochures

Brochures are the cadillacs of printed material used in connection with media presentations. They typically contain a great deal of graphic and photographic information, are generally printed in color and are used primarily in support of programs designed to sell a product or an idea.

DECIDING WHEN TO USE PRINT TO SUPPORT A MEDIA PROGRAM

Before deciding whether or not to supplement a media program with printed materials, try answering the following key questions:

- Does the program have goals that conflict or interfere with each other?
- Does the scope of the project include a great deal of technical information that would dilute the message?
- Would it help if the audience could have a "hard copy" of the presentation to take away with them?
- Are there diagrams or technical illustrations included in the program that cannot be accommodated by the medium?
- Are there elements of the program that the audience will need to know and remember long after attending the presentation (i.e., standards of conduct, facts and figures, etc.)?

By answering these questions you can begin to determine if a printed supplement is justified for your program. You can probably think of many more variables that would influence your decision. To better illustrate how printed materials can improve a media presentation, let's look at some examples of how they fit in with the media commonly used by business communicators.

Example: Printed Material to Support a Fundraising Project

Each year, our media productions group is enlisted to produce a 10-minute fundraising film for the company's employee-sponsored charity. The purpose of the film is to encourage employees to donate to the charity through payroll deductions. The program typically focuses on a needy group or individual who has received funds from the charity and shows employees how their contributions have a positive impact on less fortunate people in the community.

One year, the board of directors decided that the film should also include details of exactly how funds were distributed to the local community. On the surface this seemed like a reasonable request that could be easily integrated into the film. After some research, however, we discovered that the charity's allocation process was very complicated and would require more time to explain than we could spare in a 10-minute program. The additional information would also detract from the human interest message and lessen its impact. Ultimately, we realized that we were faced with a more basic problem. The primary purpose of the film was to motivate employees to contribute; adding the new section would add a second purpose to the program—informing the audience about how the charity distributed funding. After careful consideration, we decided that both objectives would be compromised if they were included in the same program.

The solution we presented, which was accepted by the board, allowed us to fulfill both requirements without compromising either message. The film focused on the human interest theme as originally planned and its primary goal of motivating the employees remained the same. A pamphlet detailing the fund distribution process and listing the organizations receiving donations was distributed to employees after the film was presented, to be reviewed at their convenience. The package was well-received and the campaign was a success.

Example: Printed Material to Supplement an Employee Indoctrination Program

Our safety department requested a videotaped indoctrination program for employees required to work in the company's hazardous material buildings. (Approximately 20 hazardous material buildings were grouped in one general area and access was carefully controlled by the plant security department.) The tape was to address general safety precautions and procedures applicable to all employees working within the controlled zone. As part of a larger training program, both new and long-term employees would be required to view the program and take a test before working in these hazardous material buildings.

The sole purpose of the program was training, so there were no problems with competing or conflicting objectives. The difficulty was the scope of the subject matter. While there were well-established general safety precautions and procedures for working with hazardous materials, each building also had its own specific set of guidelines. These guidelines were unique to each building and the nature of the materials handled within it.

We could easily produce an effective program addressing the general safety topics, but including information specific to each building would make the program unacceptably long. Also, individual employees generally worked in only one or two of 20 buildings within the zone. Giving them safety training on all of the buildings would only confuse the issue and render the entire program ineffective.

To accommodate the quantity and variety of information that was covered in the scope of the project, we decided to supplement the program with printed materials.

The videotape was designed to cover the general safety information that applied to all hazardous material buildings in the controlled zone. Separate safety pamphlets were developed for each building. Each pamphlet contained information specific to a building plus a review of the general safety material covered in the videotape. During the training session, employees were given the pamphlets for the buildings they were assigned to work in. If an employee was transferred to another building, he was provided with the pamphlet for that building prior to starting work.

Supplementing the video program with printed materials helped us to accomplish two goals. First, we were able to address both the building-specific safety information and the general safety information within the same training session. Second, employees received a hard copy of the information to take with them for later review and reference.

Example: Printed Materials to Support an Overhead Transparency or Slide Presentation

Our research and development group sponsors an annual conference that draws engineers from around the country. The two-day conference features 30 or more presentations utilizing 35mm slides or overhead transparencies. The presentations are very technical and often include dozens of visuals containing facts, figures and tables of scientific data.

At past conferences we had noticed that members of the audience often asked the presenters for hard copies of the presentations. This was so common that one year we decided to prepare a hard copy of each presentation in advance, assemble all of the presentations into a briefing book and provide it to members of the audience at the outset of the conference. This proved so successful that briefing books are now considered an essential part of all technical briefings on plantsite.

At many companies, all higher-level management presentations include both visuals (either slides or overhead transparencies) and a briefing book. The briefing book is distributed to the audience at the beginning of the presentation. Each page of the book contains the same information that appears on the visual used by the speaker. In fact, both the visual and printed versions can be produced from the same original artwork. As the speaker advances a slide or changes the overhead transparency, the audience follows along in the briefing book by turning the page. Listeners can take notes and make comments about the presentation right in the book. After the presentation the attendees keep the briefing books for later reference.

Example: Printed Materials to Support a Multi-image Program

Nearly every large company has some form of orientation program for new employees. Often, these presentations cover a large body of material, ranging from employee benefit programs to standards of employee conduct. A program of this nature could easily become an epic if some restraint is not exercised in its design.

Like many corporate media producers, we once had an opportunity to produce a new employee orientation program. The format for our presentation was to be multi-image. We settled on this medium in order to take advantage of its powerful motivational potential. After working closely with the personnel department, we determined that the program would be produced in three parts. The first part would provide an overview of the parent corporation and its internal structure. The second part would specifically address our facility and its operation; the third would be a general discussion of the details of employment (i.e., benefits, standards of conduct, pay procedures, etc.). The multi-image shows would be just part of a larger orientation program that lasted an entire day. Speaker-supported presentations by plant officials filled out the remainder of the orientation package.

Fortunately, the personnel department was way ahead of us in the area of printed supplements for the multi-image program. They had no intention of relying on the media package to ensure that new employees were familiar with the required subject matter. In addition to the media program and the speaker-supported programs, employees received printed materials outlining benefits, standards of conduct, health and safety information, corporate annual reports, and a myriad of other company related issues. Though most of this was covered to some degree in the multi-image presentation, it was important for employees to be able to refer to the printed material when in need of this information at a later time.

CONCLUSION

Not every media program can benefit from a printed supplement. Often it is not practical or possible to include a brochure, a pamphlet, or even a handout with your program. Single media presentations are often more than adequate to meet the communications challenge. However, printed materials can help solve many complex communications problems that are too large, complex, or multi-faceted for a single media presentation to handle alone. By combining printed materials with an effective media presentation, we can add a new dimension to our programs and satisfy even the most difficult communications challenge.

In our first six chapters we have reviewed media production processes. Now it's time to consider media formats most commonly used in business and industry, beginning with speaker-supported presentations.

7
Speaker-Supported Presentations

The administration of our facility changed hands recently and a new CEO took over. When the opportunity presented itself we were anxious to do the best job we could to impress the new boss. Our first assignment was to produce the annual state-of-the-business meeting for all managers at our facility. We had produced this program several times before and it had always been an extravaganza featuring a multi-image slide presentation followed by an address from the president and a question-and-answer session for the audience. This style of presentation had been an excellent match for our past president, who preferred to address his audience with a very well thought out speech written beforehand. He was an adequate, but not inspired, public speaker so the multi-image program added the entertainment and motivational elements that might have been lacking if he had relied only on his speech to communicate his message.

When the time came to meet with the new president and present our recommendations, we took the same approach and added even more pizzazz to impress him with our capabilities and creativity. But the new president wasn't at all impressed. He had a completely different presentational style and preferred talking to his audience personally, off the top of his head. He was not fond of flashy media programs, which he felt hampered his ability to get to know the people in his audience. He liked working with the house lights up so that he could tell how the people were responding to his message and delivery.

The program we finally produced, considerably different from our original proposal, consisted of a single tray of 35mm slides controlled by the president with a remote control (much as you would do if you were showing your friends slides of your latest vacation). The design of the slides was very simple, white words on a pale blue background. And the program was a complete success. Why? Because, like the program we produced for the prior president, it was a good match between the style of the speaker and the presentation medium. Our new president just wouldn't have been comfortable with a flashy multi-image program or videotape production. His personality and style demanded a less complicated but equally effective communications approach.

The speaker-supported presentation, a basic and straightforward approach to business communications, is the most common audiovisual format used in business and industry today. Although multi-image, film and videotape are occasionally used, the visual media most speakers choose to support their programs are 35mm slides and overhead transparencies.

Like any audiovisual medium, speaker-supported presentations have their advantages and disadvantages. Your decision whether or not to use this approach will be determined by a variety of factors, including budget and schedule as well as the speaker's style. But do remember that the speaker-supported presentation is a dynamic, valuable communications tool. It should not be overlooked in favor of another medium simply because it is not as flashy or glamorous to produce.

ADVANTAGES OF SPEAKER-SUPPORTED PRESENTATIONS

Speaker-supported presentations offer the following advantages to the business communicator:

- minimal production time
- minimal production costs
- presentation flexibility
- audience interaction
- facts and figures

Let's look at each of these advantages in detail.

Minimal Production Time

Business presentations are often needed fast. It is not unusual to have a production schedule of hours or minutes rather than days and weeks. It is also common for changes to be made in presentation content right up to the time the presentation is to be given. These conditions require a visual medium that can be both produced and modified very quickly. Speaker-supported presentations are often the only kind that can be produced within the allotted time while meeting quality standards.

The reasons for this adaptability are fairly simple: first, a speaker supported presentation does not require the production of a soundtrack, since the speaker provides the narration; second, the visuals used for speaker-supported presentations are generally less sophisticated and can be produced in a relatively short period of time (as we will discuss later in this chapter); and finally, the script for speaker-supported presentations is minimal. Often there is no formal script at all. Presenters of speaker-supported materials are usually expert in the subject matter and prefer to speak extemporaneously, tailoring their remarks to the specific interests of their audience.

Minimal Production Costs

Just as business presentations must be produced quickly, they must also be cost-effective. A single high-level board meeting might include dozens of presentations from subordinate managers or sales representatives. In the course of a week, literally hundreds of presentations will be made at a large manufacturing plant. Supporting these presentations economically requires a cost-effective presentation medium.

Production costs for speaker-supported presentations are generally lower than those for film, video, multi-image, and slide-tape programs because there are fewer components to be created. In fact, as we'll discuss later in this chapter, basic speaker-supported presentations can be produced with only a typewriter and a copier. Additionally, the equipment required for speaker-supported presentations is inexpensive and simple to operate.

Presentation Flexibility

Business communicators often find it necessary to modify the content or design of a presentation to suit a specific audience. This can mean going into more or less detail, adding new material or rearranging a program to alter its message. This can be complicated with film or videotape productions since they generally include a soundtrack and are edited into complete presentation packages. Changing or modifying these programs usually requires re-editing, which is both costly and time consuming.

Speaker-supported presentations are easily modified to fit the needs of the presenter and the audience at minimal cost in time and money. With no soundtrack to modify, the speaker simply adjusts his verbal presentation to reflect the changes he desires. Visuals can be replaced, removed or rearranged without a costly re-edit of the program. And, as we stated earlier, visuals are quick and inexpensive to produce. Speaker-supported presentations provide the business communicator with a dynamic medium that can be easily adapted to meet changing presentation requirements.

Audience Interaction

Speaker-supported presentations offer the presenter a real opportunity to receive feedback from the audience *during* the presentation. This can come in the form of question-and-answer sessions or in casual ongoing discussions. Speaker-supported presentations are often less formal than other media and thus encourage active participation by the audience. Based on audience response, a presenter can even modify a presentation during the program to meet the needs of the audience.

Facts and Figures

Speaker-supported presentations are most effective for presenting objective, straightforward subject matter such as facts and figures and are a good choice for informing and educating an audience. In the business and industrial field, examples of this usage would include a presentation to upper management on sales figures for the last quarter, a departmental status report on productivity, and a meeting to inform employees of new policies and procedures. The goal of a speaker-supported presentation should be to provide the audience with information they want or need to know.

DISADVANTAGES OF SPEAKER-SUPPORTED PRESENTATIONS

Although the speaker-supported presentation offers many advantages to the business communicator, it is not the proper medium for every communications need. The following disadvantages must be considered before deciding on a speaker-supported presentation:

- inconsistency,
- diminished impact
- reduced "entertainment" value.

Inconsistency

Some of the very factors that make speaker-supported presentations desirable for one application make them undesirable for another. Their innate inconsistency is one such factor. Because the speaker-supported presentation relies on a live speaker to present information, the material will be presented in a slightly different manner each time it is delivered. This can be an asset if the goal is to tailor each presentation to each specific audience. However, if the presentation of a consistent, reliable message is important, the speaker-supported presentation is not the best possible choice. This is especially true if more than one person will be making the presentation. Each individual will deliver a speaker-supported presentation in his or her own style. This can be a real disadvantage if the program is designed to provide the audience with a consistent corporate image or identity.

Here is an example from our own experience. We were asked to develop a new public tour program for a government facility. The goal of the program was to provide visitors with the consistent positive message that the facility was a good neighbor. The program stressed the company's commitment to protecting the health, safety and environment of the surrounding community. The tour program we were to replace was a speaker-supported presentation that was being delivered by several different presenters. The presenters were all plant employees who brought to the situation different viewpoints, experiences, specialties, etc. As a result, the presentations were considerably different each time. Individual presenters emphasized the areas they were most experienced in and most comfortable addressing. In many cases, the presentations took a direction contrary to the program goals.

Our solution was to produce a new tour program with a "canned," or prerecorded, soundtrack. This allowed the facility to present a consistent message to all plant visitors without diluting the message with varying interpretations of the subject matter.

Diminished Impact

While film, video and multi-image presentations can incorporate narration, music, sound effects and motion to increase program impact, speaker-supported presentations are limited to static visuals, the skill of the speaker and the information contained in the program to achieve the desired results.

Another factor that reduces the impact of many speaker-supported presentations is the location of equipment used during the presentation. Most speaker-supported presentations are made in small conference rooms that are not equipped with an adjoining projection booth. Overhead and slide projectors are placed in the conference room with the speaker and the audience. The resulting logistical distractions can reduce the effectiveness of the presentation.

For example, if the presentation is made with overhead transparencies, the overhead projector will be located at the front of the room next to the speaker for easy access. During the presentation the speaker changes the visuals by hand as he makes his presentation. As a result, the audience tends to spend much of the time watching the presenter go through these motions. The fan for the projector will be running and this will detract from what is being said. The light for the overhead projector is very bright and this can be distracting for those who must look beyond this light to see the visual as it is projected onto the screen. Such distractions reduce the visual impact of a presentation made with overhead transparencies.

If the presentation is made with 35mm slides, the distractions are lessened to some degree, since slide projectors can be placed near the back of the room. This minimizes the distraction caused by bright projection bulbs and slides can be changed by remote control rather than manually. However, because most presentations occur in relatively small rooms, noise from the motor and the turning of the slide carousel can reduce the impact of the program. The presenter must compete with all such distractions for the audience's attention.

Non-Entertaining

Speaker-supported presentations generally fall short if the goal is to entertain the audience. We say "generally" because there is an exception to this rule. A very skillful presenter can turn a speaker-supported presentation into a very entertaining experience. However, most business presenters are much more adept at performing their jobs than they are at public speaking. For this reason, speaker-supported presentations in business and industry tend to be rather straightforward and more informational and educational in nature. The pacing and delivery of a speaker-supported presentation are driven by the speaker rather than the soundtrack and movement on the screen. Visuals tend to be "text heavy," relying mostly on facts, figures, charts and graphs to communicate the message.

This is not to say that the audience will not be interested in the information being presented. Presumably, the presentation will contain information the audience will want or need to know. However, anyone who routinely sits in on speaker-supported presentations for business knows there is a considerable difference between an interesting program and an entertaining one.

PRODUCING A SPEAKER-SUPPORTED PRESENTATION

There are a number of factors to consider when producing a speaker-supported presentation.

- Purpose (Why is the presentation being made?)
- Budget (How big or how little?)
- Schedule (When is it going to be made?)
- Location (Where will it be made?)
- Format (35mm slides or overhead transparencies?)

If you have time to complete the project analysis we reviewed in Chapter 1, these questions will all be answered. However, speaker-supported presentations typically have to be put together fast and often a brief conversation with the client or speaker will have to suffice. Be sure to take notes and record all of the information you will need to make production decisions.

After your conversation with the client you should know why the presentation is being made and to whom; how much money has been set aside for production; and the presentation date and location. You should then be able to determine the format of the visuals you will be producing.

Normally, the next production step would be researching and writing the script. However, as we stated earlier, most speaker-supported presentations do not utilize a script in the formal sense. Usually, the speaker or producer creates a kind of storyboard showing exactly what he wants on each visual. This is in the form of text, rough drawings of charts or graphs, or descriptions of photographs to be used. This storyboard is then broken down into its component parts and distributed to the appropriate professionals for production.

Production

Some overhead transparency programs are produced with only a typewriter and a copy machine. However, most presentations are somewhat more involved and require the services of professionals in several fields. If you work in a large corporation, these professionals are most likely available to you through a media services department at your facility. If you do not have access to these services at your place of work, there are many small companies and individuals that specialize in the preparation of visual materials for presentations; consult your Yellow Pages. Let's take a look at some of the individuals you may work with when putting together a speaker-supported presentation.

Media Producer

The media producer coordinates all aspects of production from the initial request through program completion. The producer expedites the flow of information and materials between the artists and technicians who will be working on the project. If the presentation is produced through a media production company or department, a media producer will be assigned to the project. In a corporate environment it is not unusual for the speaker or client to perform these duties.

Artist

A skilled graphic artist is invaluable to a producer. The artist transforms your rough drafts of text and drawings into graphics and charts, advises you on color selection and warns you if you are trying to put too much information on a single chart.

Typesetter

The typesetter takes your copy and re-types it on a modern typesetting machine. You will have to specify the type style (or "font") and size desired. This type will then go to "paste up," where the artist assembles the artwork for your visual materials. Often the artist will have a preferred typesetter and will handle this task for you.

Photographer

If your presentation includes photographs, you will need to seek out the services of a professional photographer. Recent advances in photographic technology, such as super automatic 35mm SLRs, have made it tempting to forego the services of a professional and take the pictures yourself. We would strongly advise against this. The professional possesses not only the latest in equipment, but also a deeper understanding of what makes photographs effective as communications tools. We've all heard the saying, "A picture is worth a thousand words." This is especially true in business presentations; but only if it's a good picture.

Graphic Arts Specialist

The graphic arts specialist will perform several functions for the producer, from photographing the "camera ready" art to produce transparencies, to coordinating production of printed copies of the presentation in the form of books or handouts if applicable. Most print shops have all of the equipment and personnel necessary to provide these services. Some of these duties may be carried out by the photographer, depending upon how work is delegated.

It is important to note that many audiovisual production houses have all of the above services under one roof. This allows you to go to one place to contract for all of the production services you may require. This can be a great time and money saver. But be sure to shop around because prices can vary widely.

Production Options

Speaker-supported presentations vary from simple to so complex that they incorporate a multitude of production techniques. In order to illustrate the variety of options available to producers, we will walk you through several examples, ranging from a very basic program to one involving many sophisticated production techniques.

Example: Overhead Transparencies

As we mentioned earlier, an overhead transparency presentation can be produced using only a typewriter and a copy machine. Aside from a blackboard and chalk or handwritten flip charts, this is the quickest and most basic form of speaker-supported presentation. Production can be completed without calling on outside professionals and, although it does not result in a particularly attractive visual, it does meet the needs of many presentations requiring a fast turnaround time. Here's what you have to do.

Type the presentation charts on 8 1/2" bond paper, each chart on a single piece of paper placed in the typewriter horizontally. If the typewriter has interchangeable type wheels or balls, select a large, bold type face. Center and balance the words on the page so they are easy to read and visually appealing. Once all of the charts are typed, use an office copier to produce your visuals. Most copiers on the market today will reproduce an image on a clear or colored

78 MEDIA FOR BUSINESS

piece of transparent acetate. (There are several brands on the market; check with the manufacturer of your copier to determine which acetate product is recommended for your machine.) Load the copier with the appropriate acetate sheets and operate the same as for a standard paper copy. An alternative to the office copier is the 3M Transparency Maker (see Figure 7.1), a tabletop unit for producing projectable transparencies. Either method results in a projectable reproduction of the typed copy and can be used on virtually any overhead transparency projector.

Naturally, this production method has its limitations. Visuals should be limited to text (i.e., word charts or graphs). Although typical office copiers do a fair job on photographs that have been reproduced in print for publications such as newspapers, they do not reproduce photographic prints very well. Multi-color charts are only possible if your office is equipped with a copier capable of color reproduction. And finally, the overall image quality will be acceptable only for informal presentations. Don't use this production method if quality visuals are essential to the presentation.

Figure 7.1: 3M transparency maker

Example: A More Professional Look

In our second example we'll look at a technique that results in a more professional look and can be used for either overhead transparencies or 35mm slides. As in the first example, we'll be dealing with a final product that is black and white, i.e., black (opaque) letters on a clear (transparent) film. We'll also talk about incorporating photographs into the presentation.

As in the first example, we start with the rough storyboard outline of what is to appear on each visual. A typed storyboard is preferable to a handwritten one to help eliminate communication breakdowns due to poor or sloppy penmanship. Remember, the storyboard may contain text (word charts or graphs) or photographs. To make the process more understandable, we'll follow the production of the text first, and then we'll look at the production of the paragraphs.

After the storyboard and photographs are submitted to the artist who will be working on the project, the artist and the producer make design decisions concerning production of the camera-ready artwork. They will determine, among other things, the type face and type size for each graphic, page layouts, and the best chart styles for the information being illustrated. Once these decisions are made, the copy is given to the typesetter, who creates a hard copy or photographic image of the text. This hard copy is then returned to the artist for production of the graphics. Taking into account the storyboard and the producer's wishes, the artist cuts and pastes this copy onto a page and adds any original artwork needed to complement the text. (Original artwork is used when producing charts and graphs or to illustrate concepts or principles that cannot be easily photographed.)

When the artist is finished, a single page of camera-ready art will exist for each graphic. Once this is proofed and approved by the speaker or client, it is turned over to the graphic arts specialist.

The graphic arts specialist converts the camera-ready art to overhead transparencies or 35mm slides. There is a variety of photographic methods available for accomplishing this task, depending upon the desired results. We'll examine one of the more common methods.

To produce a visual that has black (opaque) letters on a clear background, the graphic arts specialist photographs the camera-ready artwork with a camera designed for reproducing flat copy. Since the artist typically pastes up the artwork with black letters on white paper, the graphic arts specialist photographs the artwork with a high-contrast, *positive* imaging film. This means that the image on the film will be the same as that on the artwork; it will not be reversed as it would be if *negative* imaging film were used. The high-contrast film sees anything in the image area as either black (opaque) or white (transparent). This film is not capable of reproducing grays.

For overhead transparencies, a single piece of 8 1/2" x 11" film is used to photograph each piece of artwork. Slides are photographed with 35mm film in a 35mm camera. Each piece of artwork is photographed and the film is processed. The individual pieces of 8 1/2" film are dried and mounted in overhead transparency frames; the 35mm slides are mounted in slide frames. The visuals are now ready to be used in the presentation.

The same basic process is followed to reproduce photographs. The graphic arts specialist takes the photographs provided by the producer/speaker and photographs them with the same camera used for the charts and graphs. However, in this case, a *continuous tone* film is used instead of the high-contrast media. Continuous tone film is capable of reproducing black, white, and all shades of gray in between. The result is a projectable transparency of the black and white photograph.

Example: Color Visuals

In this example we'll explore production techniques for color visuals. There are numerous techniques available and new methods are continually being developed. We'll briefly look at two of the production options currently available and discuss the advantages and disadvantages of each.

Reflected Copy

In this established technique, the visuals for text and graphics are created or pasted up using colored paper, pens, paints or other color media. Original artwork, paintings, drawings, or virtually any type of artwork can be incorporated into the presentation. This artwork is then photographed on a copy stand using a color transparency film. Existing color photographs can also be rephotographed in the same way. An advantage of this method is its accessibility; most in-house corporate media departments have all the equipment necessary to produce this kind of presentation visual. The drawback is the time needed to produce the visual. This is a labor-intensive production method, so the cost is higher and the turnaround time longer.

Computer Graphics

Computer-generated graphics are rapidly becoming a primary means of producing color charts and graphs for speaker-supported presentations. With this method, the artist creates the graphic with the aid of a computer. The color graphic is displayed on the computer screen as the artist designs the image. To create the graphic the artist uses a variety of tools, including the computer keyboard, a mouse or puck to create freehand drawings and charts and a variety of software packages designed specifically for producing presentation graphics. Computers used for this purpose range from relatively common IBM and Apple units, to highly sophisticated, specialized computers such as the Genegraphic models (see Figure 7.2).

Computer graphics vary from relatively simple color word charts to extremely complex visuals using a variety of design elements, including photographs. The only limitations are the imagination of the artist and the program budget. Naturally, a presentation featuring the more sophisticated computer graphics will require a larger budget for production. However, basic computer

Figure 7.2: Genegraphics computer graphics station

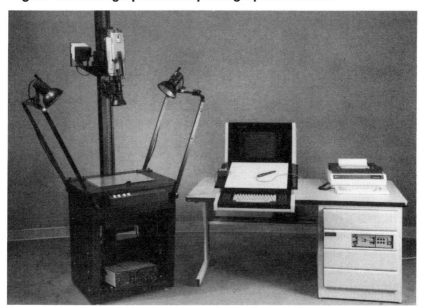

Courtesy of Genegraphics.

graphics produced on the less expensive equipment can be very competitive in price with many more traditional production techniques that are more labor-intensive.

How the actual visual you use in your presentation is produced depends upon the type of computer used to create the graphic. On smaller computers such as those sold by IBM, a camera hood or film recorder is mounted over the computer's video display screen and a photograph is taken of the image. The film is then processed, mounted, and used in the presentation.

With the more sophisticated graphics computers, such as the Genegraphic models, the image is not photographed directly from the computer's video display screen. These devices use a special high-resolution video display screen coupled with a pin-registered camera to record the image on film. Both the high-resolution video display screen and the camera are housed in a separate unit that is activated by the graphics computer.

Desktop Publishing

Desktop or electronic publishing systems are rapidly changing the way many corporations produce their internal and external publications. Speaker-supported presentations are also being produced faster and cheaper with this new technology. (For more information on desktop publishing, refer to Chapter 5, Using Graphics.)

Desktop publishing systems come in all shapes and sizes and are available from major computer distributors including IBM, Apple, AT&T and Xerox. Each system has its unique characteristics, but they all share the same basic components: computer, output device and software.

The term computer refers to the hardware, such as the IBM PC or the Apple Macintosh. The output device is the part of the system that images the printed material or the visual for the presentation. There are many types of output devices available; the more common ones include dot matrix printers, ink jet printers and laser printers. Software is what makes each system truly unique. The software is loaded into the computer (usually from floppy disk) and acts as a road map, guiding the user through the steps required to create images. Many different software packages are available and each has its own uses and applications.

Desktop publishing systems combine these three components into an efficient unit for producing presentation documents and visuals. Graphics, charts, text, and even freehand drawings can be created on the computer and then imaged as either a transparency for projection or as hard copy for distribution. All of this can be accomplished at one workstation by one operator in far less time than other, more traditional methods. It's not surprising that nearly all major corporations have begun to use desktop publishing systems because of the considerable savings in time and money.

PRESENTATION EQUIPMENT AND TECHNIQUES

Even the best transparencies and slides will fail to deliver your message to the audience if the mechanics of showing them distract from the presentation itself. Projection equipment should be reliable and appropriate for the specific setting, and the presenter should be trained in, and comfortable with, its operation.

Showing Overhead Transparencies

Overhead transparency projectors are basic in most major corporate conference rooms (see Figure 7.3). A flip of a switch is all that's usually required to start the machine and the presentation. Some models are portable and fold up into something that resembles a briefcase (see Figure 7.4). These are nice to have if you are making a presentation at an unfamiliar location and cannot confirm that a projector will be available upon your arrival. These portable models perform just as well as their larger counterparts. Other models are designed specifically for longer projection distances in auditorium settings.

Before the presentation, the speaker should turn on the projector to make sure that the unit and the bulb are working. The projector should also be focused at this time. The projector is either placed at the front of the room facing the screen or to the side of the room, the latter to lessen the distraction from the projector lamp. The speaker stands beside the projector facing the audience, with the transparencies stacked in order on a table next to the projector and within easy reach. The speaker places the first transparency on the projector and begins the presentation. Each transparency is then removed and replaced by the next visual until the presentation is completed.

Showing 35mm Slides

The 35mm slide projector is nearly as common as the overhead transparency projector and is also found in most business conference rooms. Kodak

84   MEDIA FOR BUSINESS

Figure 7.3: Bell & Howell overhead transparency projector

Figure 7.4: Briefcase overhead transparency projector; open and closed

Speaker-Supported Presentations 85

projectors are so common they have become the industry standard. The two types of professional Kodak projector you may encounter—the Ektagraphic II and the Ektagraphic III—each use the same carousel and operate in about the same way.

Using the 35mm projector is nearly as simple as using the overhead transparency projector; we'll assume it's a Kodak for purposes of this example. First, load the 35mm slides into a slide carousel that will fit on top of the projector and hold the slides during the presentation (See figure 7.5). The slots in the carousel are numbered. The first slide goes into slot 1, the second into slot 2, and so on. Each slide is placed in the carousel upside down and backwards. The carousel is then loaded onto the projector and the first slide is advanced into position. Project the slides before presentation to make sure they are correctly loaded.

The 35mm projector is placed at or near the back of the conference room. Most projectors come equipped with a zoom lens. Prior to the presentation, the image size should be checked. Make the image large enough so that your slides can be read easily from the front and back of the room. Ektagraphic projectors

Figure 7.5: Loading carousel onto a 35mm slide projector

come with a remote control so that the projector can be operated from a distance. Operate the projector with the remote control to ensure that the projector will advance, reverse and focus. In addition, make arrangements for someone in the audience to dim the lights and turn on the projector when you're ready to begin. By following these guidelines you'll ensure a professional presentation.

Computer-Assisted Presentations

A number of presentation systems currently on the market utilize a computer as the presentation tool instead of a slide or overhead projector. These systems are composed of two parts: the software, which is used to create the graphics, and the presentation device that displays the graphics on a computer screen.

With these systems, the graphics are created on a personal computer using software specially designed for business communicators. The operator enters the appropriate text or data points, specifies the type of chart required and indicates the desired colors and the computer creates a professional-looking full-color graphic. When all the visuals for a presentation are complete, they are stored on a computer disk. The disk is inserted into the presentation unit, which is attached to a computer screen. Using a remote control, the speaker can advance through the graphics one at a time, much like a slide show without a slide projector. Most systems allow a variety of transitions between visuals; dissolves, wipes and mosaics are commonly used. The visuals can be shown in order or the presenter can randomly access any visual in the presentation by punching in the appropriate number on the remote control. These systems provide speakers with a great deal of personal control over their presentations both during production and presentation. As more companies discover the flexibility these systems offer, they are sure to become more popular as a business communications tool.

Speaker-supported presentations are not the most glamorous media to produce for business communications. They are, however, the method of choice for many applications. When properly used, they are very effective tools for the business communicator.

8
Slide/Cassette Programs

Let's begin with the following exchange between a corporate manager and an in-house media producer.

> Manager: "I want a first-class video. You know, something that promotes our product... a video we can show to customers at their offices or at trade shows. I've got some ideas about content, but I'll let you take care of the creative and technical end. In fact, you can use our equipment. We just bought a portable camcorder; it's the best on the market. But as I said, I want a professional program... like the quality commercials I see on TV—only longer. I've got $2000 to spend and I need the video for a trade show in three weeks. Can you get the job done?"
>
> Producer: "No."

Our manager wouldn't even suspect that his producer's answer is not based on in-house capabilities, but on plain reality. Unfortunately, this scenario occurs all too often in business and industry, where requests for videotape programs often come from individuals who know little about the cost and time involved in producing a videotape.

In this chapter we will examine slide/cassette programs, an option that might meet this manager's needs. As we will see, producing such a program also serves as an excellent exercise for the novice producer by providing an opportunity to experiment with the basic steps necessary for more difficult productions. In addition, slide/cassette programs offer a low-cost production alternative and, if properly produced, are effective training and educational tools and augment point-of-purchase sales.

A slide/cassette program consists of a single tray of slides shown in conjunction with a preprogrammed audiotaped narration. This format provides several advantages, reviewed below:

88 MEDIA FOR BUSINESS

- Slides are easy to update or change.
- Production costs are less expensive than for film, videotape or multi-image programs.
- It presents information in a concise and uniform manner.
- It is relatively easy to distribute, store and transport.
- It can be used in individual, small group or large group situations.
- It is excellent for training, educational and point-of-purchase programs.
- Slide/cassette programs are easily repeated.

Slide/cassette programs do, of course, have some disadvantages:

- They are not flashy.
- They are rarely useful for motivational purposes.
- They cannot show motion.
- They do not operate easily in reverse or fast forward modes.
- They allow for little interaction between presenter and audience.

PRODUCING A SLIDE/CASSETTE PROGRAM

Keeping these factors in mind, here is a step-by-step guide to producing a slide/cassette program. Bear in mind that slide/cassette programs require at least four production elements—scripting, photography, audio recording and synchronization (sync) pulsing, each of which occurs at different times during the production process.

Planning

As with any other audiovisual program, you must first determine the purpose of your program and who the audience will be. The next step is to set your goals and develop specific steps that will help you achieve them. (See Chapter 1, Effective Presentation/Planning.) Let's say, for instance, that the purpose of your program is informational—to discuss new insurance benefits and filing procedures at your company. Your audience is composed of hourly employees. The purpose of the program is to describe the new benefits and inform hourly employees about new filing procedures. Note that your program should not attempt to train employees how to file for insurance benefits, but should inform them that new benefits and filing procedures are in effect.

Slide/Cassette Programs 89

At this stage, it's also important to plan for equipment needs. If you do not have equipment for programming and playing your show, you will need to rent or purchase it. Start investigating your equipment options. Contact a local audiovisual sales house or consult with a media professional for advice on equipment. Then determine which brand meets your needs and falls within your budget limitations.

Budgeting

The budgeting process for any type of production may determine whether or not the program will be funded and produced. For this reason, the budget requires special attention. Since the script is the major factor in determining costs, it must be referred to when drawing up a budget. Items called for in the script—such as special effects slides, music or computer-generated graphics—will affect your budget. For slide/cassette programs, the following items must be considered for budgeting purposes (see Figure 2.1). Note that any one or all of the items may be provided by in-house production services free of charge:

- scriptwriting fee (if outside scriptwriter is used)
- film purchase, processing and photographer fees
- audio production costs
- graphics
- projection equipment purchase or rental
- synchronization/playback equipment purchase or rental
- miscellaneous supplies and expenditures
- salaries
- contingencies

Script Fees

To determine if and when you need an outside scriptwriter, refer to Chapter 3, Scripting. For now, let's examine the costs associated with scriptwriting.

Scriptwriters charge by a number of different methods: price per finished minute, percent of budget, up-front bid; pay-in-three-parts contract; post-charge, what-the-market-will-bear; or a flat fee plus residuals.[1] Some

---

[1] Matrazzo, Donna. *The Corporate Scriptwriting Book,* Media Concepts Press, Inc., Philadelphia, PA, 1980.

scriptwriters also charge hourly rates. The most common types of script fees are the price per finished minute and hourly rate. Costs for the former range from $50 to $200 per finished program minute. If your writer charges $100 per finished minute, your total script cost for a five-minute program is $500. Hourly rates vary from $25 to $60 per hour. Figure no more than $1200 for a five-minute program at $40 per hour. (Remember that prices are different for each market. A scriptwriter in Boston may not charge the same rate as one in Dallas).

Film Costs and Photography Fees

To estimate film costs, determine the number of visuals (excluding graphics, slide duplicates and copy slides) in your program, then estimate a shooting ratio. (A shooting ratio is the ratio of the number of slides shot to the number of slides used. Many slides are not used because of under- or over-exposure, inappropriate shot angle, poor color quality, bad facial expressions, poor composition or other flaws that detract from the shot.) Shooting ratios vary from 6:1 to 20:1. Ask your photographer what his or her shooting ratio is and calculate accordingly. Multiply the number of rolls of film used by any processing fee. For instance, let's say your program calls for 60 photos and your photographer's shooting ratio is 10:1. This calls for the photographer to take approximately 600 shots, which can be accomplished with about 17 rolls of 36-exposure film. Multiply the number of rolls by the purchase and processing cost per roll; this will provide you with an estimate of your film costs.

Photographers' rates differ according to the assignment. Most assess daily fees ranging from $300 to $1500. No matter where you are located, you will probably be able to find an abundance of qualified photographers.

Fees for duplicate slides and copy stand slides are other costs you may incur. Determine how many of each you need, then contact a professional photo house to see how much they charge for these services. Average cost for a duplicate slide is about a dollar. Costs for copy stand slides vary according to the complexity, size and number. If you own or have access to duplicating equipment and/or a photo copy stand, your only costs will be film and processing.

Audio and Graphics

Audio production costs vary depending on the complexity and number of elements called for in the soundtrack. Typical production charges include the cost of the narrator, music, sound effects, tape stock and studio time. Since each audio studio has diverse capabilities, rates will vary. You can compare facilities by touring them and obtaining their rate cards. Our discussion of audio (later in this chapter and in Chapter 4) contains specific pricing information for narrators, music and sound effects.

Costs for graphics also depend on their complexity and the time it takes to produce them. Simple word slides and pre-existing art are inexpensive. Stylized titles and detailed illustrations can break the budget. Whether produced manually or generated by computer, each graphic has a production cost unto itself. Explore different graphic techniques and individual costs with an artist or designer before choosing the appropriate graphics for your show and budget.

Equipment

You need specialized equipment to program and play back your show. (Consult the section on Presentation and Equipment Techniques in this chapter for more details on equipment needs.) Such equipment can be purchased or rented from most audiovisual equipment companies or catalogs. If neither of these options is viable, contact equipment manufacturers directly. They will sell you the equipment or refer you to an authorized dealer. Most reputable dealers have no qualms about providing names and phone numbers of satisfied customers. Remember two words of advice—caveat emptor.

Supplies and Miscellaneous Equipment

During the course of production, miscellaneous supply costs can add up. Don't forget to budget for items like slide trays, glass slide mounts and labeling materials. It may also be necessary to calculate travel expenses, lodging and meals. These costs can be minimal or extraordinarily high, but failure to include them in the budget results in cost overruns that you may have to pay.

The last budget item to consider is contingencies. Budgeting for audiovisual productions is not an exact science. Therefore, contingencies help defer cost increases, cost overruns and other unforeseen expenses. This figure should be 10% to 15% percent of your total budget.

We have not covered all possible costs (e.g., costumes, models, special effects slides, etc.) associated with production of a slide/cassette program, but if you allow for the above items you will be able to estimate the majority of your budget.

Scriptwriting

The scriptwriting process for slide/cassette programs is almost identical to that for other media, including film, videotape and multi-image. (Refer to Chapter 3, Scripting, for more specific information concerning scripts.) After determining your goals and objectives but before you begin shooting photos and recording sound, you need to develop a script. First, decide who will write it—you, an in-house writer or a hired professional.

If you require the services of an independent scriptwriter, be sure to interview a few before selecting one. Ask for samples of scripts for programs they have written. Be sure the scriptwriter you hire is capable of writing nonbroadcast programs for business and industry. Many writers in the broadcast industry charge excessive rates and do not fully understand how to write for the nonbroadcast market.

If you are writing your own script, keep these suggestions in mind: 1) Be aware of the number of visuals used in relation to the show length. Don't keep a slide on the screen for more than 15 seconds; 2) It is not necessary for the visuals to duplicate the narration, provide some visual interest and variety; and 3) Write the way people speak. Formal-sounding dialog is unrealistic and does not hold audience attention.

After the research and treatment phases of the script have been completed, the writer must select a script format for the rough draft. The most commonly used script format for slide/cassette programs is the split page. Each visual to appear in the program is drawn or described in words. Get the rough draft approved by the client before undertaking the final draft. Some scripts may require second, third and even fourth drafts prior to the final draft.

The last scripting step is to prepare the final draft. During this phase, writers incorporate any changes made in the rough draft. They also finalize all visual and sound elements.

PRODUCTION PLANNING

Once the script is done, production begins. Before starting, you need to decide how to plan, schedule and produce your program. This is the preproduction or planning stage. Consider the following:

- Who will shoot the photos, one of your photographers or a freelancer?
- Will you shoot original slides or use existing slides? Do you need to make duplicates?

Slide/Cassette Programs 93

- Will you use music or sound effects in addition to narration? Can you mix the audio in-house or do you need to use outside audio production facilities?
- Does the script call for graphics? Who will produce them and how?
- What other miscellaneous materials or services will you need?

After answering these questions, you're ready to produce. But where do you begin?

Let's start with the photographic aspects of the script.

Photography

The visuals in your program will carry your show. Audiences will remember the visuals long after they have forgotten the narrator's words. By following a few simple guidelines, you can ensure interesting visuals. First, include a mix of close-ups, medium shots and long shots. Use interior and exterior shots. This adds interest to the program and helps orient the viewer. Stick with simple, bold visuals and always use horizontally shot slides. For more tips on basic photographic principles, consult one of the photography books listed in the recommended reading section of this book.

Before shooting any visuals, plan a shot list based on the script. This list details the number, location and description of each shot. Next, determine how you will obtain each shot. All the slides in your program do not have to be original. One option is to duplicate existing slides that come from other sources. We have seen entire programs comprised of images from "the company slide file." With the use of a photo copy stand, you can make slides from photographs. This is particularly effective when your program's subject is historical in nature. However, be sure not to reproduce anything which has a copyright or trademark unless you receive permission to do so.

Next, decide who will shoot the slides. In most instances, your choice of photographers will be limited to an in-house person, an outside professional or yourself. Budget, personnel availability and scheduling usually determine who will shoot the photography for your program.

Professional photographers possess a variety of camera bodies, lenses and accessories and they usually also have flash systems or standalone lighting equipment. Before selecting a photographer, ask to see some of his or her work or contact former clients.

Film processing is an important step that many first-time producers neglect. Before making a final decision about where you will take your film, make sure to see a developed test roll. Check it for consistent color, chemical streaks and the presence of any foreign matter. If the test roll looks good, then chances are the processing lab will be reliable.

Photographs are not the only visual element of slide/cassette programs. Graphics integrated with photographs increase the visual interest of slide/cassette programs.

Graphics

Graphics can show details and views of objects that photographs are incapable of showing. They are particularly useful for illustrating an abstract point. For example, a graphic would more clearly illustrate the principle of internal combustion than would a slide of a car engine. In addition, graphics communicate by incorporating words.

Generally, graphics consist of titles, illustrations, cartoons and word slides. Trained artists and illustrators usually produce them, but anyone can create pleasing, attractive graphics by following these guidelines: 1) Keep graphics, especially charts and graphs, simple, too much information confuses, overwhelms or bores audiences; 2) Use bold, uncomplicated letters, legibility is as important as content; and 3) Be aware of proper contrast. For example, don't put black letters on a dark blue background. It will be hard for the audience to see your image. (For more information see Chapter 5, Using Graphics.)

Graphics can be produced manually or by computer. When cartoons, illustrations, titles and words are hand drawn, they give programs a stylized look. Through the use of transfer type and type machines (see Figure 8.1), anyone can produce professional quality lettering. Type can also be purchased from local printers and typesetters. More difficult graphic techniques include air brushing or the use of paints and water colors. Whatever method you choose, be careful not to overuse a graphic style. The use of too many cartoons or illustrations can detract from a program and give it a cluttered unprofessional look. And the same is true of type styles. Choose lettering appropriate to your program. For example, don't use delicate letters in a program about racing tires.

However manual graphics are produced, they must be converted to 35mm slides. To do this, graphics are photographed on a copy stand using a 35mm camera and macro lens (see Figure 8.2). This relatively simple process greatly augments the visual portion of any slide/cassette program.

Recently, computer hardware and software have become important tools for the illustrator and artist. Although the hardware is limited to a few manu-

Slide/Cassette Programs 95

Figure 8.1: Kroy type machine

Figure 8.2: Photo copystand

Courtesy of Bencher, Inc.

96   MEDIA FOR BUSINESS

Figure 8.3: GPG turnkey computer graphic system

Courtesy of Genegraphics.

facturers, there are literally hundreds of computer software packages that can be used to create word slides, illustrations and even custom drawings. The simpler and less expensive programs operate on personal computers. These computer graphics programs are capable of producing charts, graphs, tables, word slides and a limited number of symbols. Usually, the data must be transmitted by modem to an imaging center where the graphic is converted to slide format. The more sophisticated (and expensive) graphics systems include work stations, film recorders and sophisticated peripherals. These systems are usually owned by private production houses or housed at corporate facilities.

Audio

Audio is another important production element. Most slide/cassette programs use one or any combination of three types of audio: narration, music and sound effects. (For more specific information about these audio elements, refer to Chapter 4, Audio).

After assembling and recording the sound elements, they are mixed onto a master tape. This final audio mix, which combines all the sound elements onto a single tape, can be done in-house or at a professional audio studio.

The audio portion of slide/cassette programs differs from other media in that each program requires an audible or inaudible synchronization pulse. This pulse indicates when to change a slide and is programmed onto the audiocassette tape.

Assembling and Programming the Show

After completing the audio track and the visuals, start to assemble the program. Using a light table, choose your visuals based on proper exposure, composition and visual content. Project each slide to check for flaws before making a final choice. Also, be certain that the visuals correspond to the shots called for in the script.

One suggestion concerning slides: use glass mounts. They are more expensive than plastic and cardboard mounts, but they will be less likely to jam in the projector. They will also keep the slide film clean and help preserve it.

By this time, you should have purchased or rented the equipment you will use for programming and playing your show. This equipment consists of a slide projector, projection lens, slide tray and cassette recorder/player capable of recording an audible or inaudible synchronization pulse.

Don't neglect the choice of projector and lens. These two items help determine the quality of your projected visuals. Be sure to choose a high-quality projector; many manufacturers make quality models. When it comes to a projection lens, there are two types of lenses from which to choose—curved or flat field. Use curved field lenses when your slides are mounted in cardboard or plastic. Use flat field lenses when your slides are mounted in glass.

There are other options to consider when choosing a lens. You can opt for a fixed focal length or zoom lens. The fixed focal length projects one image size, while zoom lenses allow you to change the image size while maintaining the same distance to the screen.

There are only two types of slide trays you need to consider, both circular. The 80-slide capacity tray is preferable to the 140-slide tray. The former is less likely to cause slides to jam in the projector. The latter, although it holds more, is more likely to cause problems. In addition, it cannot accommodate glass mounts.

There is a variety of slide-sync cassette recorder/players. They differ from consumer cassette decks in that they can record a synchronization pulse. However, the synchronization pulse generated by one piece of equipment is not necessarily compatible with that of other different brands. If your program is to be shown on equipment different from yours, be sure the sync pulses are the same. You can purchase or rent this equipment separately or as a combined unit. Whatever equipment you use, read the manual and become familiar with the operation before you begin programming.

Presentation Techniques and Equipment

A slide projector and an audiocassette deck with synchronization playback capability are the main pieces of equipment you'll need to play your show once it is produced, but the planning process doesn't stop here.

Many excellent programs have been ruined by omitting these final steps. To ensure a professional presentation, play the show from start to finish *before* the audience arrives. Adjust the sound level for the room. Check the visuals. Watch for improperly mounted, upside down or backwards slides. Be sure the lens projects a large enough image. If possible, use an auto-focus projector and replace the old projector bulb with a new one. Finally, try to keep power cords away from aisles, entrances and exits.

Once the audience is seated, it is helpful to have a speaker introduce the program and explain why it is being shown. When the speaker finishes, dim or turn off the lights and start the program. If you have planned and rehearsed the presentation, all should go well. However, problems can occur and you need to be ready for them. Slides will jam, projector bulbs will blow and audiotape will break. Your only recourse is to turn on the lights and correct the problem. Don't be embarrassed. Most audiences will sympathize with your misfortune. They will not be as understanding, however, if you don't know how to extract the jammed slide, have not brought a spare bulb or don't have an extra copy of your audiotape on hand. Don't look like an amateur; and don't let a good program be ruined by neglecting common sense presentation techniques.

9
Multi-Image

Let's start this chapter with a conversation one of our friends related to us:

Client: I need a program to kick off my annual sales meeting and motivate 200 people to sell more. I was thinking of film or videotape.

Producer: I'd suggest a medium many of your salespeople may have never seen before. It's flashy *and* motivational.

Client: What is it?

Producer: Multi-image.

Client: You mean a slide show?

Producer: No. I mean multi-image.

As this dialog suggests, multi-image is a relatively new presentation format. It combines computer technology with two or more slide projectors and an audio track to create a unique presentational medium. Simply defined, this consists of multiple images projected onto a specific screen format and programmed to a prerecorded sound track. Note the difference between multi-image and multi-media. Multi-image combines both visual and audio elements into one medium. Multi-media, on the other hand, is the use of several media in one presentation or setting. A multi-media presentation may consist of film, multi-image and videotape programs. It can also include lasers, guest celebrities, choreographed dancers and other media forms.

When properly produced, multi-image programs are entertaining, flashy and stimulating. And, since few people have seen a multi-image program, the format can surprise and captivate audiences. For these reasons, some producers prefer this medium.

Multi-image programs are used to introduce new products and services; they are shown as motivational pieces at sales conventions; they sometimes serve as the "opening act" for training and educational programs; and their value as informational tools is unlimited.

Here are some examples of how multi-image programs are employed. Many ski equipment and clothing companies premiere their new lines with the assistance of six-projector multi-image shows. The medium lends itself well to the excitement and fashion of the sport. Many large companies (including Chrysler, Avon and Coca-Cola) incorporate extravagant multi-image programs that employ 20 projectors or more as one element of their annual sales convention. The programs often act as inspirational pieces that summarize the company's accomplishments and help motivate salespeople for the coming year.

Multi-image programs are also excellent for introducing educational and training topics. These smaller programs, utilizing two to nine projectors, can provide an entertaining overview and focus attention on the subjects to be taught. Week-long training sessions and orientations for newly hired employees are energized with multi-image shows. Some multi-image programs are produced solely for art and entertainment purposes and they can be simple or quite complex. They present information in a thought-provoking manner, while engaging the audience in a visual phantasmagoria. Examples of these types of programs include those shown at Disneyland, Epcot Center, and in major art museums and IMAX facilities. Other tourist attractions like Churchill Downs, New York's waterfront and Vancouver's historical district feature informative multi-image programs.

PRE-PRODUCTION CONSIDERATIONS

Multi-image is most effective as a motivational tool. Since the quality of multi-image visuals (the contrast, grain and sharpness of 35mm slides) and sound tracks is much superior to both 16mm film and videotape, this is the medium for delivering powerful messages. Multi-image programs have some distinct advantages over other media:

- They contain extremely high-quality visuals and audio tracks.
- They can be used in individual, small or large group settings.

- They are regarded as unique by many audiences unfamiliar with this medium.
- They create an exciting and lasting impression.
- It is fairly easy to repeat multi-image programs.

But before we go on, it is important to point out that multi-image productions have some severe drawbacks; more than most other media.

- Production techniques can be complex, costly and time consuming.
- It is difficult to update or change multi-image programs.
- Because of its complexity, multi-image equipment is often prone to technical problems.
- Multi-image programs and the associated equipment are difficult to distribute, store and transport.
- There is little compatibility among equipment manufacturers.

Assuming the advantages outweigh the disadvantages and you decide to produce a multi-image program, here is how to go about it.

Planning

Like most other audiovisual programs, multi-image presentations require a detailed project analysis, all the more important given multi-image's complexity. For example, the fact that multi-image programs can be shown in many different screen configurations (see Figure 9.1) means there is an additional planning step not necessary for other media programs (including motion pictures and videotape, which use only one projection format). As with other media, you need to determine a multi-image program's specific purpose, goals, objectives and audience. (For more information about project analysis, refer to Chapter 1—Effective Presentation Planning.)

Multi-image is a specialized medium, and although many corporations have in-house multi-image capabilities, independent studios still produce the majority of multi-image programs. For this reason, we suggest that you contact a reputable production house if you require a multi-image production. An experienced producer will establish a concept, supervise production of the program and even purchase playback equipment. Or you may choose to work with individual specialists—scriptwriters, photographers, directors, graphic artists, etc.—to help you with each phase of your program.

If you choose to go the production house route, shop around. Ask to see programs the company has completed. Contact previous clients for their input.

Figure 9.1: Screen configuration examples

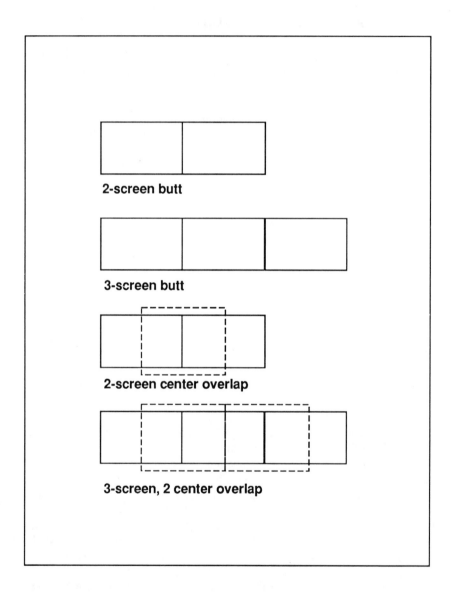

Once you select a producer, give that person as much information and help as possible. As the executive producer, you have as much invested in the final product as the producer and you need to keep the lines of communication open.

If you plan to purchase multi-image equipment, this is the time to start investigating alternatives. Your needs will vary according to the complexity of the program and where it is to be shown. All multi-image programs require the following equipment (see Figure 9.2):

- slide projectors and lenses
- projector stands
- programming and/or dissolve units
- audio equipment
- projection screen
- miscellaneous cords and other equipment

Budgeting

Because multi-image production vastly differs from other media production, it presents a greater budget challenge. With adequate planning, the budgeting process can be as concise as for any other medium. The budget items for a multi-image program are similar to those for a slide/cassette program (see Chapter 8, Slide/Cassette Programs). However, multi-image productions use more projectors and peripheral equipment than slide/cassette programs. This, in turn, means greater costs for services and materials—programming time, photographers' fees, film, trays, slide mounts, etc. Therefore, you should expect to spend more for the production of a multi-image program. Experienced producers will provide you with a detailed budget estimate. Although costs may seem high, working with an experienced person makes it more likely that you will waste little time and resources and end up with a quality, professional program.

With or without a producer, you will need budget line estimates for some or all of the following items for a multi-image program:

- scriptwriting fee
- photographer fees and stock photo charges
- 35mm film purchase and processing
- audio production costs
- graphics production costs
- programming of the show

Figure 9.2: Multi-image equipment: Projector, reel to reel recorder/player and programer

- programming, projection and playback equipment purchase or rental
- materials and supplies
- contingencies

The Script

For everything you need to know about budgeting for scripts (what to look for in a scriptwriter, how they charge and what you can expect to pay, please refer to Chapter 3, Scripting. We'll simply note here that writers are generally paid a percentage of the total budget, by the hour or on a per finished minute basis.

The Photographer, Film and Processing

Audiences remember visuals long after they have forgotten the key words in a multi-image program, so take the time you need to choose the right photographer and don't be afraid to be selective. This is a big budget item; professional multi-image photographers normally charge high daily fees because they are more specialized than normal photographers. (In Denver, for example, daily charges range from $400 to $1200.) Be sure to discuss fees, either with your producer or directly with the photographer. Before hiring a photographer, examine the person's portfolio. If you like what you see, then the photographer's rate is appropriate.

If you cannot afford the services of an award-winning multi-image photographer, you have two other options: hire a professional photographer who is unaccustomed to shooting for multi-image programs or use an in-house photographer whose salary is paid by the company. Make sure the person you select is familiar with the differences between regular photography and multi-image photography. (For example, all multi-image photos must be shot in a horizontal format or they will not conform to normal screen dimensions. Also, images must be large, bold and simple. Subtle photos do not work well in multi-image.)

The other budgeting concerns for photography are film purchase and processing. We discuss this process in detail in Chapter 8, Slide/Cassette Programs. Professional photographers, whether in-house or freelance, will recommend the type of film for use in your program. Before photography begins, have your photographer determine a shooting ratio (the number of slides taken compared to the number used). This will help you estimate film and processing costs.

Stock Photos and Graphics

Some scripts call for visuals that cannot be easily obtained or require great expense. Let's suppose you need shots of the Egyptian pyramids. Obviously, the expense of flying a photographer to Egypt to shoot the pyramids would be prohibitive for most budgets. Stock photo houses, located in most major metropolitan areas, offer a solution to this problem. Stock photo houses have thousands of slides on file of foreign countries, one-time events, seasonal shots, historical events, generic industrial shots, and a wide variety of others. The fee for use of a specific shot is based on a formula. The elements used to calculate your cost include type of medium in which you will use the visual, number of times the program will be shown, the length of time you will use the visual and various other factors. After the fee is determined, either you duplicate the slide you wish to use or the stock photo house duplicates it for you. Prices generally range from $30 to $200 depending on the specific visuals and the number you select.

Graphics are another type of visual used in multi-image programs. Individual costs are determined by the number of graphic elements, the time it takes to produce them and the types of equipment and materials used in the production process. Low-budget word slides produced on a computer or by an artist can cost as little as $10 each. More stylized graphics usually take longer to produce, which means higher production costs. (This is true for both manually and computer-generated graphics.)

Graphics can be classified as follows: title slides, credit slides, bulleted word slides, charts and graphs, illustrations, tables, freehand drawings, and others. Most often, these graphics will be superimposed on a colored or patterned background that adds to the aesthetics of the visual.

Title slides appear at the beginning of the program and may also occur at the start of individual sections within the show. Credit slides usually appear at the end of the program and list the people or companies that played a major role in the production. As in Hollywood films, the most important roles are those of executive producer, producer, director and the various production roles—photographer, narrator, scriptwriter and the like.

Bulleted word slides, another option for displaying information in a multi-image program, consist of a title and a list of items which appear underneath. For instance, a bulleted word slide may have the title TYPES OF AUDIO. Listed underneath on separate lines with a bullet would be Music, Spoken Word, Sound Effects, Silence and Natural Sounds.

Charts and graphs, illustrations, tables, cartoons and animation are all discussed in Chapter 5, Graphics for Media Presentations. When producing graphics, be sure to avoid the common pitfalls outlined in that chapter.

Audio

Audio production charges include the cost of the narrator(s), music, sound effects, tape stock, studio time and miscellaneous materials, as well as field recordings such as interviews or other sounds. Whatever the budget limitations, don't treat audio like a poor relative. Audio technology in multi-image programs offers a distinct advantage over other media. Whereas television and film sound are often played through small speakers, multi-image productions most often have their sound tracks played through sophisticated, high-quality sound systems. This quality difference allows multi-image programs to sound more professional. Don't overlook their potential.

Programming

Programming equipment is an integral part of any multi-image production (see Figure 9.2). Through the programmer, a producer manipulates and transforms the images the audience sees on the screen. Although this looks like magic to the uninitiated audience, it is actually accomplished by programming equipment that consists of a dedicated computer and dissolve unit(s) that control the rate at which slide projector lamps turn on and off. The programming equipment generates "cues," which are referred to as a cut or dissolve (the temporary overlapping of two images whereby one visual is replaced with another). These cues range in time from 1/10 second to 64 seconds.

Here's how it works. When you depress keys on a programming keypad or keyboard, the programmer generates a discrete electronic pulse for each specific cue. The dissolve unit receives this data, then transfers the cue information to the slide projector. The projector performs the function (a cut, dissolve, flash, etc.) when it receives the cue. As this occurs, an audiotape player/recorder (see Figure 9.2) simultaneously plays the prerecorded sound track while recording the cues on a separate audio track. This preserves the cues for future playback.

Programming equipment also creates special effects such as flashing, blinking, freezes and repetitions, which are useful for animation. In addition, programming equipment controls the forward or backward movement of individual slide trays.

As we mentioned above, you must decide whether to rent or purchase programming equipment from the five major brands of programming equipment currently available: Audio Visual Laboratories (AVL), which dominates the U.S. market; Clear Light, Arion; Multivision; and Electrosonic. Unlike videotape and film, there is little compatibility among brands. Although manufacturers are attempting to rectify this problem, competition still stands in the way of industrywide standardization. For this reason, be sure to select the proper system for your needs. Programming capability and the number of projectors a programmer can handle are the main factors that determine price. The more complex the programmer, the more expensive it will be. Units that can program three projectors start at about $900 and go as high as $2000. More sophisticated programming equipment controls upwards of 30 to 100 projectors. Prices are commensurate with capability.

Contingency Items

Producers too often fail to budget for extraneous materials and supplies. Costs for glass slide mounts, projector trays, graphics supplies and alignment slides can add up, especially if you are producing a 15-projector program. As a fellow producer friend says, "Be sure to calculate every single expense you will incur. If you don't, the small stuff will put you over budget." So don't forget contingencies. This 10% to 15% cushion defers cost increases and other unforeseen expenses. And since multi-image budgeting can be an inexact science, the contingency line item is a necessity.

THE PRODUCTION PROCESS

Now that your budget is set, we'll outline the production process, emphasizing factors unique to multi-image presentations.

Scriptwriting

Scriptwriters who specialize in multi-image programs are rare and there are few standardized methods for writing multi-image scripts. The most popular script formats are the split page and storyboard. You'll also have the choice of words only (for which producers provide a visual design and style), words and visual descriptions, and treatment outlines with visual design ideas. Regardless of the format your scriptwriter uses, be sure it is clear, visually oriented and written to be heard, not read. As Palmer E. Dyer states, "In multi-image programming, the visuals are supplied and the primary emphasis must be on what appears before the viewer—not on what may be heard from the narrative accompaniment."[1]

If you must hire a scriptwriter, be sure to see script and program samples. Also ask experienced producers who they use. Our advice is to hire a professional producer who will then hire a scriptwriter with whom he or she has worked.

Photography

Visuals carry a multi-image program. As mentioned above, multi-image photography differs from regular photography. (For more general information on photography, refer to Chapter 8, Slide/Cassette Programs). Here are a few differences:

- Multi-image photos cannot be viewed for extended periods of time because slides are constantly changing on the screen.
- Slides should always be shot in a horizontal format.
- Shoot large, bold, simple subjects.
- Shoot a good mix of long, medium and close-up shots.
- Multi-image photography often involves special effects.

The producer usually accompanies the photographer during the shoot so he or she can provide photographic direction. If you give the photographer a shot list, which describes the type of visuals to shoot, give him or her creative license to shoot any other visuals that look interesting.

Another photographic consideration involves graphics; not all graphics are imaged to slides. Some may have to be shot to slide on a photo copy stand (see Figure 9.3). Other existing slides may have to be duplicated. These are uncomplicated processes, but be sure to include these shots in your production schedule.

---

[1]Gordon, Roger L., Dyer Palmer E., *The Art of Multi-Image.* Abington, PA: Association for Multi-Image International, 1983.

Figure 9.3: Using a Bencher photo copy stand

Special visual effects are often called for in multi-image programs. Common effects include glows, animation, panoramas, split images and types of movement such as streaks and zooms. These effects are a combination of visual production and programming technique. First, the producer must determine if the programming equipment can perform the special effect. If the special effect is feasible, it requires precise planning in the photographic stage. Before shooting, the multi-image producer and photographer meet to review the exact composition, subject placement, lighting and other visual elements that must appear in the slide. Our advice is to confer with an expert from a professional slide services company or a special effects photographer for specific questions concerning special visual effects.

Audio

Audio is an important element of any multi-image show and helps determine the success or failure of the program. Because of its importance, some producers devote as much time and energy to the sound portion of the shown as they do to the visuals. In Chapter 4, we provide a detailed analysis of each sound element, its advantages, disadvantages and cost.

Like other media programs, multi-image productions often have large budgets and are produced for special occasions. For this reason, celebrity narrators are often used. But more important than narrator fame are voice tone, inflection, timbre, pace and other qualities. These traits determine how your narrator will sound and whether that person can convincingly deliver the spoken part of the program. You can find narrators through talent agencies, the Screen Actors Guild (SAG), radio stations, and other sources. For details on this process as well as selection of music and sound effects, see Chapter 4, Audio.

Staging

Staging, or showing the program, is the final step in the multi-image process. The person in charge of staging may be the client, the producer or a technical expert who is hired to set up and show the program. Regardless of who assumes this responsibility, any person involved in staging should keep this checklist in mind:

- the date(s) and time(s) for program presentation
- the location for the program presentation

- a brief description of the show
- the length of the program
- the number of screens used and their arrangement or layout
- type of projection (front or rear)
- the number and type of projectors necessary
- the number and focal length size of projection lenses
- the type and brand of programming equipment needed for play back of the program
- audio requirements
- special staging effects

A few of these items require clarification. Let's start with projection. Projection of multi-image programs occurs from either the front or the rear. In front projection, the slide projectors are typically placed in an enclosed booth area to prevent the audience from hearing the sound of the operating equipment. Slides are projected through a window onto a screen in front of the audience, while the sound track is played over a speaker system inside the auditorium or viewing area. Occasionally, the dissolve units, slide projectors and associated equipment are placed in the same room as the audience. Although the clacking sound of changing slides may distract the audience, this situation is unavoidable in most hotel and conference room settings.

In rear projection, the multi-image program and equipment are situated behind a rear-projection screen. No equipment is visible to the audience. The slides must be reversed when projected on rear screens in order to correctly display the visual information. The projection method has a direct bearing on the type and size of lenses that need to be used to show the program. Most audiovisual catalogs have charts which compute the size of lens to use based upon screen size and projection distance.

Beautifully produced sound tracks can be ruined by poor playback systems. For this reason, staging people need to determine audio playback requirements. The size of the room in which the program will be shown and the number of people in the audience help determine the appropriate type of audio system. In-house auditoriums usually have adequate sound systems. Other facilities, such as hotels, rarely host multi-image programs and their sound systems are designed only for amplification of speeches. Therefore, audio systems must be rented to accommodate multi-image sound tracks. Other items—program length and description—are valuable to the staging person and help in the rehearsal phase. The staging crew rarely get the opportunity to see the programs they will project. If they know how a program begins and ends, and its length, they can identify problems and troubleshoot them before the audience sees the program. This saves time and allows your program to be shown in a professional manner.

Transfer to Videotape

Since multi-image equipment is cumbersome and requires technical expertise to operate, many programs are transferred to videotape. Although the videotaped copy simplifies program play back, the result is a smaller, grainier, higher-contrast image. The sound track is usually heard through a three-inch speaker, not a sophisticated sound system. Consequently, the program's impact is reduced and it does not affect the audience in the same manner as originally intended. But due to cost and ease of transport, the transfer to videotape is gaining acceptance and popularity.

If your multi-image program will be transferred to videotape, design the program with that in mind. The following are some basic design rules to use when multi-image transfer to videotape is likely:

- Design big, bold simple visuals. Words and other graphics that seem large on screen will look smaller on a television.
- Avoid high-contrast visuals. They will appear grainy on videotape.
- Do not use black and white together. Television cameras have difficulty reproducing these film tones.
- Avoid reds; they do not transfer well to videotape.
- Not all multi-image moves work well on videotape. Quick cuts, fast animation sequences and other special effects may have to be redesigned for television.
- Be aware of the different aspect ratios of slides and video (see Chapter 5). The aspect ratio of slides is 2x3 (two units high by three units wide). Television's aspect ratio is 3x4 (three units high by four units wide).
- Make sure all visuals are shot "TV safe." For instance, all titles must appear in the safe area so that they are not cropped on the top, sides or bottom during the transfer process.

If you follow these rules, your program will transfer well to videotape.

There are two basic methods of transfer: off-the-screen and aerial imagery. Off-the-screen is accomplished by projecting the program onto a screen (front- or rear-projection), recording the image off the screen with a high-quality video camera, and running the audio track directly into the audio input of the videotape recorder. This method is inexpensive. Transfers for a three-projector program can cost as little as $300. However, this method does not replicate the image quality of the original visuals as well as does aerial imagery, which involves a sophisticated system that does not use a screen. Images are projected directly onto a video camera tube and then recorded on videotape. Aerial imagery transfer retains the high quality of your visuals and allows for video special effects moves. Unfortunately, it is expensive. Prices are based on the number of projectors and visuals, but expect to spend a few thousand dollars for a three-projector, one screen transfer. Aerial image transfers are most readily available in Los Angeles and New York and a few other large cities.

CONCLUSIONS

Multi-image is an enigmatic medium. Some argue that it is the most creative and artistic medium available to producers; others say its inherent equipment problems and lack of portability will ensure its obsolescence. We believe that multi-image has limitless possibilities. And although the mass conversion to videotape has drastically reduced the number of multi-image programs, its effectiveness as a motivational tool remains unchallenged. The medium leaves a lasting impression on its audience, as anyone who has seen a first-class, multi-image production knows. That's the real beauty and value of multi-image.

10
Film Production

Customer: "I need a program that will give our employees a sense of company history. I want them to know where we've been so they can better understand where we're going."

Media producer: "How many employees do you have?"

Customer: "About 10,000."

Media producer: "How many employees will see the program at one time?

Customer: "Between 500 and 1000; it's part of a companywide program to improve morale."

Media producer: "Are there any special requirements you can think of?"

Customer: "Only that we want it to look good, really good."

Media producer: "Sounds like this program should be produced in film."

FILM TO VIDEO: A SHORT HISTORY

The industrial use of motion pictures grew continually through the 1950s and 60s and by the mid-70s, at its peak, industrial uses of film included employee training, marketing, public relations, documentation, and virtually all communications applications. Nearly every major corporation either supported in-house film production facilities or relied on independent producers to provide a constant supply of films. The standard format became 16mm film and no conference room was complete without a motion picture projector in addition to the more traditional slide and overhead transparency projectors.

In the late 1970s, video began to challenge film's dominance as an industrial communications tool. Video technology advanced quickly and as it did, the cost of production equipment, once a prohibitive factor, dropped to within reach of the industrial communicator. Commercial television stations, which, up until then, had relied exclusively on 16mm film, began to switch to video for field reporting. Creative and technical professionals in both the commercial and industrial markets began scrambling to learn all they could about video in order to ensure their marketability. Industrial media production facilities across the nation followed the example of their commercial counterparts and began to make the switch from film to video. This trend continued and by the mid-1980s video had eclipsed 16mm film in the industrial market. Even die-hard film afficionados had to admit that video was ideally suited for most corporate communications needs.

Although this would seem to paint a dreary future for the industrial motion picture, film continues to play a role in corporate communications. Even businesses that have invested heavily in video production facilities generally maintain at least a minimal film production capability. Other companies that produce video in-house still produce film with the assistance of independent film specialists.

Advantages of Film

In spite of the many conveniences offered by video, film has a number of advantages that make it the medium of choice for certain applications.

Appearance

Much has been made of the superior "look" of film as compared to video. Motion picture film has a greater contrast range (ability to record a wide range of dark and light values). The contrast range of videotape is considerably less than that of film and the resulting image appears flatter and often less appealing. Film records an image that is closer to what the human eye actually sees, thus a program shot on film often has a more authentic "feel" than does a video image. When the budget allows, most television commercials are filmed rather than videotaped because of the greater emotional appeal of the film image. For the same reason, many high-end corporate programs designed to sell or market a product are still produced on film and then transferred to videotape for convenience of presentation.

Presentation

Film can be shown in virtually any setting to an audience of virtually any size. Suitable facilities for film presentation range from small conference rooms seating 10 viewers to large commercial theaters accommodating several thousand patrons. Anyone who has seen a feature film in a modern motion picture theater can attest to the impact of the large screen image. Although most businesses cannot maintain presentation facilities on this large scale, virtually all corporations support a number of smaller conference rooms and auditoriums seating from a dozen to several hundred employees. A small investment in modern film projection equipment (including an adequate audio system) can turn these facilities into mini-theaters ideal for film presentations. Such intimate settings can actually enhance the effectiveness of the film being viewed.

Duplication

Multiple film copies, or release prints, can be ordered from the lab in virtually any quantity and the format can vary with the presenter's needs. If the film was produced on 16mm stock, copies can be made on 8mm or 16mm film with no noticeable loss in quality. When 35mm release prints are made from a 16mm original there is some loss in quality due to the enlargement required for the larger format.

Film can also be successfully transferred to videotape. This represents a major advantage over videotape, which can only be transferred to film with major losses in visual quality. For this reason, many independent producers of industrial media produce most of their projects on film. This allows them to provide their clients with a quality program in either film or video format.

Disadvantages of Film

There are, however, some drawbacks to using film. Here are some factors to keep in mind when choosing a production format.

Reliance on Vendor Services

As we mentioned earlier, most large corporations have retained a minimal in-house film production capability consisting of a camera and audio recording equipment, a moviola or flatbed editor, and a variety of film splicing tools. While this will allow an in-house media production department to film and edit a program, there is a large body of work that must be procured from outside vendors. Film processing, audio mixing, negative conforming and duplication are just some of the operations that must be performed by a specialized film production facility. Not only is this costly, but it adds a considerable amount of time to the production schedule. Unlike film, videotape productions can be produced almost entirely in-house. Even the smallest video production facility, equipped with a camera, cuts-only editor, minimal audio mixing equipment and an inexpensive character generator, has all of the equipment necessary to produce a complete, simple videotape program from start to finish.

Availability of vendor services can be a problem if you are not located in a major media market like Los Angeles, New York or Chicago. In many smaller cities, there are few (if any) motion picture laboratories or support services. As a result, you may find it necessary to seek out these services in another city or even another state. While most film professionals are accustomed to working on projects in this manner, it does require additional time, money, and planning.

Cost

Because of its reliance on vendor services, film is becoming very expensive to produce. For this reason, film is generally being used for "high-end" productions that absolutely require the superior film look. On the positive side, as costs for film production services continue to rise, costs for industrial video production equipment are falling, and video is quickly becoming within reach of even the smallest in-house corporate media departments.

Turnaround Time

Imagine this scenario: Your company has just won a contract that will ensure its future well into the next decade. An announcement will be released to the media in two days—but your president wants the employees to know before they read it in the newspapers. He calls you requesting a media program in which he will address the troops and outline the major details of the contract. The program must be produced, duplicated, and delivered to plant managers for showing to employees prior to the release. What do you do?

This scenario may seem a bit extreme, but similar requests are made of in-house media departments every day. The turnaround time for a film project prohibits its use for this application. Filming, processing, editing and duplicating the finished program would take most corporate operations at least a week to complete. Videotape, however, could be quickly produced and edited in-house, duplicated, and released in time to meet the president's request.

As this story illustrates, the turnaround time required for film production is no longer acceptable for many corporate communications needs. Video's immediacy has changed corporate expectations; media packages have to be ready now; later just isn't good enough.

PRODUCING A FILM

While most corporate/industrial motion pictures are produced on a smaller scale than their Hollywood cousins, they all share the same basic approach. Regardless of the scope of the film, there are three basic phases in the production process: pre-production, production and post-production. Each of these phases involves a body of work that is critical to the success of the final production. The scope of this book does not allow us to provide an in-depth discussion of the film production process; this section is intended to provide a basic overview for those new to film production. If you would like to supplement the information presented here, several excellent resource books are listed in the bibliography.

Pre-production

This is the first and arguably the most critical phase of the film production process. Decisions made during pre-production will determine the success

or failure of the film project. Some of the many steps completed during pre-production include:
- the presentation analysis
- writing the script
- selecting film stock and format
- selecting a crew
- selecting performance talent
- scouting locations

The Presentation Analysis

We strongly recommend that a thorough presentation analysis be completed as the first step in the pre-production process. This will help you define critical production information such as program content, audience, budget, program goals, and the identity of your technical experts. A presentation analysis will also help to ensure that your vision of the program is similar to that of the client. A complete description of the presentation analysis is covered in Chapter 1.

Developing the Script: Research to Treatment to Script

Like overall film production, development of the motion picture script is broken down into several basic steps. The first step is researching the subject matter. Depending on the topic of the program, this may be completed at the public library, through interviews with technical experts, with reading materials provided by the client, or any combination of these and other methods. It is not uncommon for the scriptwriter to become something of an authority on the film's subject by the time the script is completed.

Following the research, the information is organized into an outline that details the content to be covered and suggests a logical order of presentation. Depending on the complexity of the program, this outline may be presented to the client for his approval.

Next, a treatment is developed. Basically, a treatment is a narrative description of how the film will look and sound. If the film will tell a story, the treatment covers the major plot points and story elements to appear in the final script. If the project is a training film, the treatment will describe the information to be presented. The treatment often includes sections of actual dialog or narration and may also include detailed scene descriptions if these are critical to the production. The treatment is typically much shorter than the actual script and more conceptual in nature. Details such as scene numbers, camera moves and lighting instructions are not included. The purpose of the treatment is to give the client a feel for how the film will appear as a final product. The treatment should be submitted to the client for approval prior to beginning the actual script.

Next, the actual script is written. The script is the blueprint the director will use to produce the film. In it, the writer fleshes out the concept as presented in the treatment. Everything that will be seen or heard in the final film is detailed in the script, so it is essential that the script be as thorough as possible. There are a variety of scripting formats used today. We believe the corporate teleplay (see Figure 3.3) is best suited for corporate/industrial applications. As you can see, this format is very similar to the one used for feature film and commercial television scripts. Its value to corporate producers is the way it encourages the client to read both the narration *and* the scene descriptions. In other formats such as the split page (see Figure 3.1) the client will tend to read only the narration and ignore the rest of the script. To avoid changes later, it is far safer to get the client's approval on the *entire* script. You should expect to prepare several drafts (one to three is about average for corporate scripts) to accommodate the client's changes. (For more information on the scriptwriting process, refer to Chapter 3.)

Selecting a Production Crew

While the script is being completed, you should also be thinking about who will actually produce your program. Assuming you will act as your own producer, you'll still need a director, a cameraperson, possibly an audio technician, and a host of other support personnel. The size of the crew and how they are selected depends largely on the budget and where the program will be produced. On a major feature film set, the production crew may consist of 30 to 50 people, each a specialist working within a clearly defined discipline. At the other end of the spectrum, the corporate production crew typically consists of only two to four people. With a crew this small, each member must be prepared to assume multiple production responsibilities. For instance, one person may serve as director, producer, and cameraperson. The rest of the crew might consist of an audio technician and a production assistant, or grip.

Since this book is titled *Media for Business,* we'll assume you aren't producing a major Hollywood feature just yet, so let's look at how a crew might be selected for a corporate film.

122 MEDIA FOR BUSINESS

If your program is to be produced in-house, the selection process might be over almost before it starts, since corporate production facilities typically have a limited staff and select their crews on the basis of availability. Don't be discouraged, however; most in-house production staffs consist of highly talented professionals capable of producing top quality products. The manager of the in-house media productions unit will determine the requirements of your program and assign a production crew to the project. The size of the crew will depend on the budget and the complexity of the project.

Another way to obtain a production crew is to contact an independent commercial production house. These facilities can usually provide a complete crew tailored to your budget and production needs. It is a good idea to compare several facilities before making your final choice. Ask to see examples of completed productions and talk to former clients. Prices can vary for the same type of service, so compare rate cards. Bear in mind that some production houses specialize in certain types of production. For example, some facilities might specialize in producing television commercials while others might concentrate on corporate/industrial clients. Others serve a wider client base consisting of both corporate and commercial accounts.

Freelance film professionals can also be called on to staff your production crew. To obtain the names of freelancers, contact professional film/video organizations active in your area. The International Television Association (ITVA) and The Society of Motion Picture and Television Engineers (SMPTE) are two such organizations. When interviewing possible crew members, feel free to request client lists and take a look at demonstration reels. One strategy when using freelance professionals is to select your director first and ask him or her to recommend a production crew. Directors often work with the same crew and their familiarity with each other can save you time and money, while helping create a better program

Equipment

A discussion of the tools of the film production trade could take up a whole book. Each phase of film production and post-production requires specialized equipment. If you are working with an in-house media production department they will have access to all the equipment required to produce your film. If you are working with freelance production personnel, rely on their advice regarding equipment. Freelance producers may have some or all of the equipment on hand depending on the size of the company you are working with. Rental is another option; most film production and post-production equipment can be rented from vendors specializing in this area.

Selecting Performance Talent

Once you have a director, it's time to begin thinking about the talent involved in your film, either on or off camera. Performance talent (narrators, actors, actresses, comedians, mimes, etc.), if properly utilized, can make an enormous contribution to the success of your program.

This is another area where the director's experience can save you time and money. Often, directors are familiar with local performers who might be well-suited to your program. A talented, experienced actor or actress will require fewer retakes and rehearsals on the set. And that saves you valuable production time.

Where do you find these actors and actresses? If you work with a talent or casting agency, be specific about describing the kind of performer you require. If you are looking for an actor, should he be tall or short? Does the role require or preclude a specific ethnic background? What age range are you looking for? If you need a narrator, be prepared to describe the type of voice you want. Don't let the personal nature of these considerations inhibit you. The agency needs to know this information in order to choose the people or person best suited to your needs.

Once the agency understands your requirements, they will set up an audition or casting session at which a number of performers who meet your basic requirements will be assembled. You will provide the agency with a sample of the script and they will provide copies to the performers. Then, one by one, the performers will act out the scene or read the narration for you. Remember that this is a first reading; the actors will make mistakes. Pay more attention to the person and less attention to his or her ability to memorize the script. You should feel free to give the performers any background that might help them to give a better reading. The director should be present to provide any coaching that might be helpful. Once you have seen all of the readings, you may wish to see several of the better candidates again. This selection process will continue until you have made a final choice.

Another option is nonprofessional talent, commonly used in low-budget productions. For instance, in-house production facilities will often identify several employees who have pleasant speaking voices and are willing to provide off-camera narration (often without compensation) for their film projects. The drawback to using nonprofessionals is the extra time and effort often required to obtain a usable performance. Nonprofessionals typically require numerous retakes and continuous moral support on the set, so it is often more economical to find the necessary funding to secure a professional.

Nonprofessional talent can be a desirable choice, regardless of budgetary considerations, when credibility is on the line. On-camera interviews with subject matter experts or a message from the company president are examples of situations where nonprofessional talent adds believability to the program. In these situations, the person's natural speech patterns and unpolished on-camera presence can actually be beneficial.

Scouting Locations

The director, cameraperson and audio technician should visit each location before production begins. There are many details to consider at each location in order to avoid unexpected problems later on. These include (but are not limited to):

Lighting conditions
What type of available light exists at the location? How many additional lights will be needed? Is electrical power available? If not, is a generator required? Will color-correcting gels be needed? Are there any large windows that must be dealt with?

Acoustics
What kind of ambient sound exists at the location? What kinds of microphones will be needed? Can noisy equipment in the area be turned off for the shoot?

Security/safety
Are there special safety considerations that would require hard hats, safety glasses or hearing protection for the crew? Will the area need to be secured? Will local authorities need to be alerted prior to the shoot? Is crowd control necessary?

Union considerations
Will local union regulations affect the shoot? Will you have to hire local support personnel (electricians, carpenters, crew members, etc.) or can your crew do the work?

Script suitability
Does the location meet the requirements of the script? Will modifications have to be made? Is the location suitable for multiple scenes?

Other considerations
Is food for the crew available nearby or will refreshments have to be brought in? If the shoot is to last several days, what accommodations are available? Are film support services available nearby? What about rental equipment?

Any one of these factors could delay or cancel a location shoot if not considered beforehand. A thorough location scout is an essential pre-production step that will speed the production process and eliminate unpleasant surprises later.

Production

Okay. You've secured a finished script, arranged for a crew, performed a thorough location scout and lined up talent to perform in your film. What's next? Production. The entire crew goes on location or to the studio and records what will eventually become your motion picture.

The Crew
Film production is a team effort and everyone on the crew has certain duties and responsibilities which we'll outline in the following section.

Director
Like any team, a film production crew has a captain; on the set, that person is the director. The director controls all aspects of the production process while shooting is taking place. He or she determines camera locations, coaches the talent on their performance, communicates with the cameraperson about the way the shot should look and coordinates all other production activities. Since film production is a creative endeavor, there will always be differing views on how to interpret the script. Most directors are willing to listen to ideas from other crew members. However, in all cases, the final decision rests with the director.

Before production begins, the director breaks the script down into a shooting schedule that lists all of the shots to be filmed on each day of production and includes a contingency plan for exterior shots that may be canceled due to weather or other unexpected problems. The shooting schedule indicates any special elements that each shot may require. For example, if a film requires that the talent be wearing a specific color hard hat in a particular scene, this will be noted in the shooting schedule so that the hat can be obtained before production begins. On most corporate/industrial films, the director will also oversee much of the post-production activities which we'll cover later in this chapter.

Cameraperson

The responsibility for recording the image on film rests with the cameraperson (or director of photography as this role is called in Hollywood). The cameraperson works very closely with the director to determine how the final image should look on the screen, so it is essential that they communicate effectively throughout the production process. The cameraperson's first duty is to choose the best film stock based on the lighting requirements unique to the production. The cameraperson also places lights for each set up and chooses filters, gels, diffusion, and other lighting accessories. The cameraperson acts as liaison between the film production unit and the film processing laboratory, supervising the processing and printing as well as any special handling.

Audio technician

Sound for motion pictures is taped on an audio recorder that runs in synchronization with the film camera, allowing the film and the audiotape to be aligned so that the action on the screen matches the sound that is heard. The person responsible for this is the audio technician. His or her duties include operating the recording equipment, monitoring the audio signal during the shoot, selecting the proper microphone for each recording application and recording any non-synchronous location or background sound that may be required to add atmosphere or credibility to a scene. Like the cameraperson, the audio tech works closely with the director. He or she also supervises the mixing and re-recording of audio during post-production and editing of the final product.

Production assistants

The production assistants provide administrative support to the director and other members of the creative staff. Their duties include copying scripts, running to the store to pick up everything from batteries to doughnuts, filling out forms and providing any help required on the set. This is one of the less glamorous positions on the crew, but many successful film professionals began their careers as production assistants. It is an excellent way to break into the business while providing valuable support to the production team.

Grips

Grips are the muscle of the production team. Cameras, lighting gear, audio equipment, dollies, props, and anything else you can imagine are handled by the grips. They also assist in placing lights for the scene, assembling dolly track and rigging electrical cables. Sometimes the camera crew, the audio crew and the lighting crew will each have a specialized team of grips familiar with their area of expertise. Unfortunately, this is a luxury few industrial filmmakers enjoy.

Now that our crew is assembled, let's move on to the next production step.

Setting Up the Shot

The basic building block of the motion picture is the shot. A shot can be of anything—an action, a line of dialog or the surrounding landscape. A series of related shots, when edited together, becomes a scene and a series of scenes becomes the finished film. Naturally, the selection of shots can't be random. They are dictated by the script, the blueprint for the film you are producing. How is a shot set up? We'll start at the beginning and review the whole process required to film a single shot for an industrial/corporate motion picture production. For purposes of illustration, we will assume that this shot will be filmed on location with professional actors.

When the director and crew arrive on location, their first task is to determine the exact camera position for the shot. This decision is based on the shot description in the script and how the director has visualized the scene in his mind. Since most single shots are part of a larger scene, the director will also be thinking about how this shot will be intercut with others. A director often shoots what is called an establishing, or master, shot first. This is typically a wide shot and features all action or dialog that will be in the final completed scene. After the master shot is completed, the director shoots any close-ups, reverse angles, reaction shots, etc., that will be necessary when editing the completed scene.

Once the camera position for the first shot is determined, work can begin. The camera is placed on its platform (either a tripod or a dolly) and secured into position. The director works with the cameraperson to determine the proper lens for the shot. The cameraperson then begins to light the shot based on how the director would like the scene to appear on film. The sound technician assembles the audio recording equipment and selects the proper microphones. If the actors will be wearing microphones on their clothing, the audio tech attaches them and adjusts the audio levels. The director works with the actors on their dialog, and coaches them on their delivery and movements (walking, motioning, etc.).

Once all is ready, the director calls for quiet and instructs the cameraperson to begin rolling film. Once the camera is rolling at the proper speed, the director instructs the audio tech to begin rolling tape. When the tape recorder has also reached the proper speed, the scene is slated with the familiar Hollywood-style clapboard and the director calls for action. The actors go through their paces and finally the director calls cut, signaling the end of the shot. Shooting continues until all shots to be filmed at the location are completed. The crew then strikes the set, removes their equipment and moves to the next location to set up for the next series of shots.

Production Administration

A single take of a single shot for a motion picture can easily consume more than 100 feet of film. The typical industrial film is composed of hundreds of shots that may require one, two, or even three takes to record properly. By the time the film is completed, tens of thousands of feet of film have been exposed. As you can imagine, keeping track of this much film requires a comprehensive administrative program.

This involves the use of camera reports during production by either the director or a production assistant (see Figure 10.1). The camera report does not call for a great deal of information. An entry is made upon the completion of each shot, in the order shots appear on the reel of film. The scene number identifies the scene in the script that is being filmed. The take number indicates how many times the scene was filmed. A check in the sound column indicates that sound was recorded with the take. The footage number is copied from a visual readout located on the camera. Finally, there is room for the director's comments on the scene. This information is very helpful when identifying camera footage during the editing process.

Another administrative task involves marking the exposed reels of film before they are sent to the laboratory for processing. At the end of each day's shooting, the cameraperson removes the exposed film from the camera magazines and places it in a sealed canister to prevent accidental opening. At this point, there are no marks on the film to identify it as yours. To prevent confusion later on, the cameraperson then marks the reel of film with a reel number, his or her name, the name of the production company and a phone number. This ensures that the film you send to the lab is the same film you get back the next day.

When the entire film has been shot and all footage is "in the can," it is time to move on to the next step in the process, post-production.

Post-production

In post-production, the picture, the sound, the special effects and all of the other production elements are manipulated and transformed into a motion picture. This is the magical phase of the process because you can actually begin to see the results of your labors for the first time. During post-production, you will be relying on a variety of support services which include, but are not limited to, the film laboratory, an audio production facility and a negative conformer. Depending upon your location, these services may all be provided by a single post-production facility. In other markets you may have to contract for each service separately. In any case, review a number of vendors before making a decision. Ask for a client list and references. Compare prices and remember to factor in extra costs if the vendor you select is not located in your immediate area.

After the original camera footage is processed at the film laboratory, a copy or workprint is made. The original is stored at the lab in a climate-controlled vault and kept there until much later in the post-production process. The workprint is a positive print and can be projected on a screen or viewed in an editing machine. When it is returned from the lab it is screened by the producer, director, cameraperson and other members of the creative staff. If the footage seems suitable and no re-shooting is necessary, it is turned over to the editor so the post-production process can begin.

But first, the sound recorded on 1/4-inch audiotape during production has to be transferred to 16mm full coat stock. This is basically a 16mm film base coated with oxide on one side (see Figure 10.2). The 1/4-inch tape is copied onto the full coat stock with an audio dubber. The full coat can then be placed on a film editing machine and played back in synchronization with the film image. For motion pictures with synchronous sound, the full coat and the workprint are edited simultaneously. Audio transfers can be performed at most film post-production facilities, including film laboratories.

Figure 10.1: Camera report

CAMERA REPORT

Date _____ Company _____
Director _____ Cameraperson _____
Production title _____
Magazine # _____ Roll # _____ Film stock ____

Scene Number	Take Number	Sound	Footage	Comments

Figure 10.2: 16mm full coat

The next step in post-production involves logging each reel of footage on a log sheet (see Figure 10.3). Like the camera report, the log sheet lists the scene and take numbers and provides room for comments about the scene. The log sheet also lists edge numbers which are located along the edge of the film next to the sprocket holes (see Figure 10.4). They appear every 20 frames (every six inches) on 16mm film and are numbered sequentially from the beginning of the reel to the end. They are photographically printed on the camera original by the manufacturer and are visible after the film is processed. The numbers that appear on the workprint are printed from the original when the copy is made. Film editors use edge numbers to identify the location of a specific shot to within 20 frames. When the entire motion picture has been completely edited, edge numbers are used by the conformer to cut the camera original to match the edited workprint. After completing a log sheet for every reel of workprint to be used in the edit, the audio is synchronized with the workprint and the reels are physically broken down shot by shot. Each take is cut out of the reel, marked, and set aside along with its matching audio. No footage is discarded at this time. Even the "bad" takes are kept and labeled. When all reels have been broken down, the footage is ready to be edited.

Figure 10.3: Log sheet

Production: _____ Producer: _____ Camera Roll # _____

Edge number	Scene#	Take #	Comments

Figure 10.4: Edge number on 16mm film

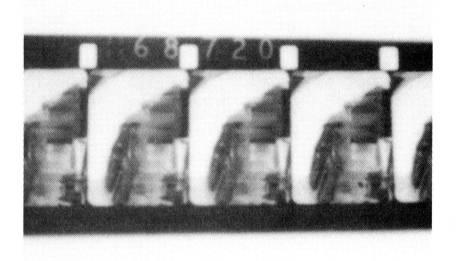

Editing

The first step in the editing process is called the assembly. Both the film and audio elements are assembled onto reels in approximately the order called for in the script. The end of one shot is spliced to the beginning of the next shot and so on until all footage has been attached in this manner. Excess footage is not trimmed out at this time, but the slates are removed from the beginning of each shot. The purpose of the assembly is simply to organize the footage to be edited. No attempt is made to match action or intercut scenes.

When the assembly is complete, the rough cut can begin. During the rough cut, unnecessary footage left in during the assembly is eliminated. Both the film footage and the audio elements are trimmed and adjusted to fit more closely together. The editor also begins to select specific shot angles, to match action, and to intercut scenes. Close-ups and cut-aways are also inserted to emphasize drama and adjust pacing. For the first time in the editing process, creative decisions are made that will determine how the film will ultimately look. Each scene may be edited and re-edited several times during the rough cut. Shots may be cut out of the film only to be retrieved and reinserted. Under no circumstances should any of the workprint be thrown away at this point.

Pieces of film as small as a single frame should be kept and maintained in case they are needed later. The rough cut is complete when all major editing decisions have been made and the film is "roughly" edited from start to finish.

Next, during the fine cut, the process of honing and polishing the film continues. During this stage, each scene is reviewed and fine-tuned until it is as good as it can be. Edit decisions made during the fine cut are less sweeping, but no less important than those made during the rough cut. The length of a scene may be slightly shortened or lengthened; a cut-away may be added or deleted; scenes with matching action are adjusted. Overall, the pace and flow of the film is manipulated until it feels correct. When each shot in each scene is finally locked down, the fine cut is complete.

The Audio Mix

Up to this point, only a portion of the audio elements has been edited with the workprint. These elements may include synchronous dialog recorded during production, voice-over narration, and music (if the film included sequences that were edited to a specific musical selection). Now that the fine cut is complete, additional audio such as background music, ambient (background) sound, and special audio effects may be edited into the motion picture. Each audio element (music, ambient sound, sound effects, dialog, etc.) is assembled onto a separate reel and edited so that it is heard at the right moment in the film. For instance, suppose one scene depicts two people walking through a meadow. You might want to insert the ambient sounds of birds chirping in the background to add atmosphere to the scene.

There are several ways to obtain the audio effects and ambient sound that you want to include in the program. Naturally, some ambient sound and special effects will have been recorded while you were on location. (If you filmed a scene that featured a highway for instance, it would have been a simple matter to record the sound of cars rushing by the location.) However, some sounds are not so easy to come by. The sound of a gun firing, for instance, does not record well on tape. This effect requires considerable enhancement in an audio studio before it will sound authentic. Fortunately, there are libraries of pre-recorded sound effects and ambient sounds available for a fee to film producers. Most professional film/sound studios have access to these libraries and can re-record the effects you require.

The editing process is complete when the final adjustments have been made to both the workprint and the audio elements. At this point, the film will consist of one reel of edited workprint (the visual image) and several reels of edited audio elements (dialog, ambient sound, narration music, etc.). The next step is post-production is the audio mix, during which the audio reels are re-recorded and blended together to make a single reel of audio that will be the soundtrack for the finished film.

The sound mix is performed at a professional film/sound studio, where each reel of audio is placed on a machine called a dubber. The edited workprint is placed on a specialized film projector designed to run in sync with the audio elements on the dubbers. Both film projectors and dubbers are controlled by an audio engineer at an audio mixing board located in the studio. The film is projected onto a screen located in the studio. Speakers in the studio play back the sound from the audio elements on the dubbers. When the film is projected into the studio, the audio engineer and the film director can hear all of the audio tracks running in sync with the projected image. During the mixing session the audio engineer will adjust the level of each audio element with a control located on the mixing board. When all levels are set correctly, the engineer will re-record and blend the audio from each dubber onto a single reel of full coat. This single reel will contain the mixed audio from all of the separate reels constructed during the edit.

Let's look at an example. Suppose the scene being mixed depicts two people having a conversation. In the background we can see a highway with cars rushing by. To make it interesting, we'll give one of the people in the scene a transistor radio that happens to be on during the conversation. Although the most important audio in this scene is the dialog from the conversation, we must also be able to hear the sound of the highway in the background and the music from the transistor radio. The conversation, the highway noise and the radio music represent three audio elements that must be controlled during the mix for this scene. Each of these elements is on a separate audio dubber and can be independently controlled by the audio engineer. First the engineer sets the audio level for the dubber playing the dialog. This will be the loudest audio in the scene. Next, the audio level will be set for the radio. We must be able to hear the music, but it cannot interfere with our ability to understand the conversation. Finally, the level for the background noise of the highway is set. We should be able to hear the cars going by, but just enough to know they are there. When all levels are set, the three tracks are re-recorded onto the single mixed reel that will form the soundtrack for the film. This process is repeated for each scene in the film.

When the mix is complete, the film will consist of one reel of edited workprint and one reel of mixed audiotape containing all of the audio information that will be heard in the film. For a 16mm film, this audio track will then be converted to an optical track, a picture of the soundtrack that can be interpreted and played back on a standard 16mm projector. The optical track will be used to print this picture on the answer and release prints of the film that will be made after the camera original has been conformed. For 35mm films, a magnetic track made directly from the soundtrack produced at the mixing session is used.

Conforming

Now it's time to retrieve the camera footage from the climate-controlled vault at the film lab where it has been stored. It can now be conformed to match the edited workprint. This is not a task to be taken lightly. Original camera footage is very delicate and represents the entire investment in the film up to this point. Extraordinary precautions should be taken to ensure that the footage is not damaged in any way.

First, obtain the services of a professional conformer. Most film laboratories either have a conformer on staff or can recommend one to you. Conformers, who are trained to handle original footage without causing damage, work in climate-controlled, dust-free environments and wear white, non-abrasive film gloves to prevent leaving fingerprints. Their job is to cut the original camera footage to match the edited workprint. To do so, the conformer uses the edge numbers located on the side of the film, next to the sprocket holes. If the film being conformed was photographed with 16mm film, the original footage will be cut into two printing reels called the A and B rolls. The A and B rolls are synchronized together and assembled into a "checkerboard" pattern; where roll A contains camera original, roll B contains opaque black leader and vice versa. A positive print of the film is made in two passes through the printer—first with the A roll, then with the B roll. The result is a complete print of the entire film. A and B rolls are used to facilitate clean cuts and dissolves between scenes. Two reels are necessary because of the small space between frames on 16mm film; splices used to connect one shot to another extend beyond this small space and are visible when projected. The A and B roll "checkerboard" technique prevents the splices from overlapping into the picture area. Unlike 16mm, 35mm film can be conformed without using A and B rolls because of the larger space between frames.

The first copy made from the A and B rolls is called the first answer print. This is the first attempt at properly balancing the color and exposure for each shot in the film. Printing a motion picture utilizes the same basic photographic principles as printing a still color photograph. The printer, or "timer" as this person is called in film terms, must arrive at the right combination of light and light filtration for each shot in the film. When the first answer print is screened, the timer and the cameraperson who photographed the film will review each shot and note any changes to be made. The first answer print may be followed by a second, third, or even fourth answer print before each scene is properly balanced.

When the answer print is approved by the entire creative staff, the film lab is given the go-ahead to produce the release prints. These are the copies of the film that will actually be turned over to the customer for use in the field. Upon completion of the release prints, the entire production process is finished.

THE FUTURE OF FILM IN BUSINESS AND INDUSTRY

The future of film in the corporate marketplace will depend on technological developments in the commercial arena. Most film industry experts agree that eventually film and video will become one medium. When this transformation will be complete, however, is the subject of some debate. Currently, the resolution and contrast range of video—even the new High Definition Television—does not approach that of film. However, advances in electronic imaging are occurring at a tremendous rate. It is just a matter of time until the video image equals the film image.

In some ways the merging of film and video technology has already begun. As we discussed earlier, many television programs and commercials are photographed on film to take advantage of its superior appearance and then transferred to video to utilize its technological advantages in post-production. The completed program shares the advantages of both mediums. So what impact will this have on corporate films and filmmakers? It is inevitable that film will continue to represent a smaller and smaller percentage of overall corporate media production. As video technology continues to advance, the use of film will continue to decline. By the turn of the century, it is likely that film will be completely replaced by video in the corporate marketplace. Corporate media producers should consider these facts when planning for the future. The lessons we have learned from film, lessons in creative story telling through the visual image, will always be useful; but the medium for distributing the story itself will continue to change.

11
Videotape Production

Videotape programs are rapidly becoming the medium of choice in the nonbroadcast world. This trend began in the late 1970s and early 1980s when videotape began replacing motion picture film. Today, educational institutions, governmental agencies, hospitals and companies of all sizes use videotapes for training, information, education, sales, marketing, motivation and public relations.

If so many companies use videotape, it must offer some distinct advantages, and indeed it does. First, most people are familiar with the videotape medium. Since virtually every person in the United States has a television, audiences are considered to be "video literate." This makes videotape an accessible medium for the audience. Two other obvious advantages of videotape are its ability to show motion and to show an event as it happens. In this sense it resembles film, but unlike film, videotape provides almost instant gratification. Motion picture film must be chemically processed and printed before viewing occurs; videotape is an electromagnetic medium that can be viewed immediately after recording.

Videotape has other advantages. It is an excellent medium for small audience viewing. One playback machine and a television monitor can easily accommodate up to 15 people. Videotapes also offer portability and since most businesses have some type of videotape player, showing a videotape program is easily accomplished. The equipment also allows for repetition, fast-forwarding and review of individual segments or entire programs.

Yet, there are some drawbacks to using videotape. Because years of TV-watching have made audiences "video literate," they have a tendency to compare all videotape programs to those on broadcast television. Although this is an unfair comparison since few nonbroadcast programs have budgets or content comparable to their broadcast brethren, clients and producers need to be aware of this fact when they choose videotape as the medium for their programs.

Production costs are another area that can be disadvantageous. Videotape production can be as expensive—or in some cases more expensive—than motion picture film. Some professional videotape programs cost as much as $4000/$5000 per finished minute. Also, the expense and difficulty associated with making changes in visuals or audio is not only costly, but sometimes impossible to do without beginning anew. That's why evaluation at the end of each major production phase is invaluable.

One problem that will dissipate in the next decade is the quality of the picture in relation to the size of the viewing audience. Because videotape's resolution, grain and color quality are inferior to film, and because videotape has a "flat" look, it does not project well. For this reason, videotape programs should not be shown to more than 25 people at one time unless additional monitors are employed. Large-screen projectors offer an alternative. However, only those that currently cost $10,000 and none produce a quality image.

Other alternatives are in the works. One is High Definition Television (HDTV), which has a much greater resolution and clarity than present-day videotape. European, Japanese and American companies continue to research and experiment with this wide screen format. But for now, HDTV is inaccessible to most companies.

Our familiarity with video can actually be a problem, as the following dialog illustrates:

Fred: Hi, Ellen. If you have time, I'd like to meet with you in the next couple of days.

Ellen: What's up, Fred?

Fred: Well, I've got a training program that starts in two weeks. I'd really like you to produce a videotape that will introduce the topic and serve as an overview of the training process.

Ellen: That's great Fred, but my staff is scheduled for the next two months. And besides, we'd need much more time than two weeks to produce a program.

Fred: You're kidding! Just what do your people do? It can't be that hard to produce a video. They do it three times a day for the local news.

Ellen: Listen, Fred. Those are much different circumstances. As I said, we can't help you now, but maybe we could help you on some other project.

Fred: How much notice do you need? A year?

For people like Fred, video has become more of an afterthought than an important part of the planning process. And yet planning is essential, as we'll discuss next.

PLANNING A VIDEOTAPE PRODUCTION

There are three interdependent phases in videotape production: pre-production, production and post-production. Inadequate planning in any phase will result in a poor program. For that reason, it is essential to define and clarify the planning elements. Producers and clients need to identify the purpose, goals and objectives of each program. Audience size and demographics must also be determined. How the program will be used and viewed are also important in the planning process. For example, if a proposed program is training-oriented, you need to know how it will be integrated into the training plan—is it the centerpiece or an adjunct? (For additional information on planning, refer to Chapter 1, Effective Planning.)

Presentation Techniques

In most cases, videotape programs are stand-alone presentations with introductions or explanations incorporated in the program. This makes video especially effective for informational, motivational and training purposes. However, under certain circumstances, it is wise to have a speaker introduce videotape programs and answer questions after the program ends. Imagine showing a program about how to fill out new insurance forms to an audience of employees who are not aware of the differences between the old and new benefits. At the very least, the audience would want to ask questions about the change in benefits, how it would affect their dependents, how much it would cost and a host of other questions relating to the subject of the program. Without a representative to explain the changes and answer questions about the forms, this program could conceivably raise more questions than it answers. If improperly presented, even the most polished, concise and insightful program cannot effectively communicate its message to an inadequately informed audience.

In the nonbroadcast world, videotape is most often used in conjunction with or as a training tool. Some studies estimate that more than 60% of the videotapes produced are training-related. In these situations, the program serves a specific purpose as the centerpiece of the training package, as an overview, or as an educational tool to be followed by some sort of testing procedure to determine the viewer's retention and understanding of the subject. A training videotape is usually introduced and explained by a speaker, then supplemented with written handouts or formal classes.

Budgeting

Many in-house video departments and commercial production houses break budgets out into three types of expenses: above-the-line, below-the-line and post-production. For our purposes, we will discuss budget items in more general terms. Although standardized forms for budgeting do exist, most companies and producers design their own. Figure 2.1 shows an example of a typical nonbroadcast budget. Most budgets usually include the following items: labor costs, purchased services, rentals and supplies.

Personnel expenses vary depending on whether you use in-house personnel or hire outside crews. In-house production staff receive fixed salaries paid by the company. Since most companies offer some sort of benefits package, this also needs to be figured into an in-house person's salary. Outside production crews constitute an additional but sometimes necessary expense. Crews normally include one or all of the following: a producer, director, videographer, engineer, sound technician, lighting technician, production assistant and a varying number of other technical consultants.

Equipment costs are another budget item. For those companies that do not own every piece of hardware needed to produce a videotape, equipment purchase and rental are important factors. Suppose that your production calls for a number of shots to be taken 20 feet above the ground. Unless you have access to an aerial lift truck, you will need to rent equipment to accomplish these shots. If your 12-part videotape series requires a wireless microphone for the on-camera narrator, you will have to determine whether to budget for purchase or rental of the mike. If you have no equipment, you will have to carefully and thoroughly investigate your options. Should you buy or rent? What type of equipment will suit your needs? What type of package deals do vendors offer? Do equipment rental packages come with crews? Do vendors offer daily, weekly and monthly rates? What about discounts? Answering all these questions and more will help determine equipment costs for your budget.

Those specialized services that the client or in-house producer cannot provide represent another significant budget area. These costs range from minuscule to hefty and fall into two categories: people and equipment. People generally means talent. This includes, but is not limited to, voice-over narrators, on-camera talent, actors and extras. Other specialized people services include make-up experts, food preparation specialists and animal trainers, to name just a few. Special effects editing and computer-generated graphics production account for the bulk of costs in the equipment area. Since the price of this equipment is prohibitive for many companies, use of this equipment and its associated services is rented at an outside post-production house. Many post-production houses offer a full complement of squeeze zooms, digital effects, 2-D and 3-D animation, and other graphics effects. We advise clients and producers to visit post-production facilities before using them. Ask for rate cards; these are invaluable when budgeting for specialized services such as editing.

Remote versus Studio Production

Videotape production takes place in a studio or at a remote (also called field) location. In their book *Small Format Television Production,* Compesi and Sherriffs distinguish between studio and field production as follows: studio productions usually make use of multiple cameras, while remote shoots use a single camera; studio productions have larger crews (often eight or more), while remote shoots have crews of two or three; studio productions are done live or taped live, while remote shoots include individual segments or shots which are edited or inserted into a larger program; studio productions have a greater amount of control over the environment, while remote locations must adapt to the situation.

Studios with multiple camera set-ups also employ a piece of equipment known as a switcher/special effects generator (see Figure 11.1). All the studio cameras are routed through the switcher. By pushing the appropriate camera button and/or moving the fader bar, the director can instantaneously change cameras by way of cuts, dissolves, fades and a variety of special effects. More elaborate field recordings may also employ a switcher for multiple camera productions. Although many of the same techniques are used in both the studio and in the field, there are distinctions. For the most part, the following discussion applies to remote production.

Figure 11.1: A switcher/special effects generator

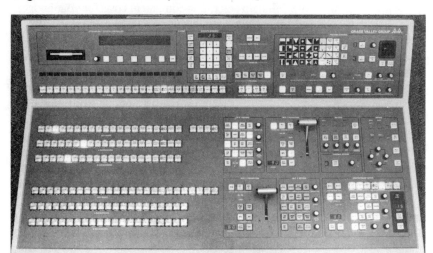

Photo courtesy of the Grass Valley Group, Inc.

PRE-PRODUCTION

Most producers and directors believe that the pre-production phase of a videotape program will ultimately determine its success or failure. In this phase, planning takes center stage and the major element to be considered is the script. After the final script has been approved, a shot list must be established. The director or someone he or she delegates is usually responsible for this. The shot list is a detailed accounting of each visual that will be needed in the final program (see Figure 11.2). The shot list may include the following items: the type of shot (original video, stock footage, graphic, etc.); the location, the time of day when the shot should occur; whether the shot is an interior or exterior; name(s) of talent to appear in the shot; the necessary props or costumes; and other important information. Transition devices in the script such as cuts, dissolves and special effects are also noted.

The producer or director must also set a shooting schedule, sometimes planned in conjunction with or as part of the shot list. This plan outlines the times and locations for each shot. It also identifies who needs to be where and

Figure 11.2: A shot list

```
                        PDC SHOT LIST
FRCC

    7  —INT-Female videographer
    9  —STOCK SHOT—HAZMATT slide
   10  —INT—Drafting
   13  —INT—MS of desktop publishing room
   39  —INT—WS of Advisory Committee
   40  —INT—2-shot of Advisory Committee
   41  —INT—CU of pencil and pad at Adv. Com.
   42  —INT—CU of Adv. Comm. member
   43  —INT—CU of another Adv. Comm. member
   44  —INT—Low angle WS of Adv. Comm.
   46  —INT—WS of Exec. Comm. in session
   52  —INT—CU of needs assessment form
   53  —INT—Vocational educator filling out needs assessment form
   54  —INT—Nina at desk evaluating needs assessment
   55  —INT—CU of Central Region brochure
   61  —INT—CU of Recredential book
   63  —INT—MS of Electronic mail system
   64  —INT—CU of regional directory of PD activitie
   66  —INT—CU of list of experts for talent pool
   74  —INT—CU of mini-grant flyer
   75  —INT—CU of funding proposal cover
   80  —INT—Voc Ed person meeting with Nina in library
   81  —INT—Voc Ed person from #80 in prof. devel. setting
   82  —INT—CU of Voc Ed person from #81
   86  —INT—Videoconference with monitor
   90  —INT—Voc. Ed. person in class
   94  —INT—Voc. Ed. person at flip chart—shot through two people seated
            (stage this shot)
   95  —INT—Same Voc. Ed. person from #94 smiling
   98  —EXT—MS of FRCC sign
  102  —EXT—LS of campus
  106  —EXT—MS of child care center

Red Rocks CC

    6  —INT—Nursing (MS of patient in wheelchair)
    7  —INT—Wastewater technologist at work
   11  —INT—MS of Computer programming
   14  —INT—MS of Accounting teacher in class
   15  —INT—CU of #14
   68  —INT—MS of Sex Equity library
   69  —INT—CU of book in hand from Sex Equity library
   93  —INT—Voc. Ed. person in library looking at materials
  103  —INT—Teacher in keypad class with business machines

CCD & MSC

    2  —INT—MCU of Graphic arts person at printer
   12  —INT—2-shot of Auto mechanics at work
```

at what time. For instance, Day #2 of the shoot may require the services of three crew people, an on-camera narrator, two employees who will be shown in the background of the scene, and a wireless microphone for the on-camera narrator. All this information would be contained in the shooting schedule. The good director always has a back-up plan. If it rains on the day the director planned to shoot exteriors, the alternative plan goes into effect.

During the pre-production phase, the crew should prepare equipment—charge batteries, check cords, clean lenses, and make sure that the necessary equipment is available and operates properly.

PRODUCTION

The production phase, also referred to as "the shoot," involves the recording of visuals and sound. This is accomplished in a variety of ways. Visuals and sound can be recorded separately or together, but adherence to technique is most important in this phase. Let's look at the first component of videotape production: the visuals.

Visuals

Visuals include live action that is videotaped, such as on-camera narrators, actors and others who perform a role in front of the camera. These people are referred to as talent, whether they are professional or amateur. If you have the time and money, we advise using professionals. They usually require fewer takes for each scene (thus speeding up the shooting) and will make your program look and sound more polished.

Other common visuals are the graphics, which consist of bars, charts, graphs, text, titles, credits, names and other information. Graphics may be used alone or in conjunction with live action. For instance, one of the most common methods of identifying a person who appears on camera is to superimpose his or her name. Other typical examples include titles at the beginning of a program and credits at the end. Whether graphics occur as simple credits, ordinary charts and graphs, or complex animated titles, they should add to the understanding of the program. Too often, they are used for show and add nothing to content.

Other types of visuals including slides and still photos are conducive to television and are frequently used. By panning, tilting or zooming the camera while shooting these stills, it is possible to simulate movement, making them more interesting to the audience.

Stock footage is another type of visual used by producers and directors. This is film, video or still pictures of specific events, people, places or times of year. Why would you need to use stock footage? Suppose your script calls for a shot of a St. Patrick's Day Parade, but you're shooting your program in November? What if you needed a shot of the Statue of Liberty, but did not have the money to fly your videographer to New York? How do you get a shot of summer in the Rockies when it is January and there is a foot of snow on the ground? Chances are a stock footage house would be able to help. These companies will send you footage of specific visuals you request for use in your program. After you decide what to use, the company assesses a fee based upon the type of footage, the amount used and the program's audience. For example, a single still photo can cost $50 or more; 30 seconds of network news footage may run more than $250. Just be sure to calculate this cost when figuring your budget because stock footage purchase can be expensive.

What makes a good visual? The answer to this question is subjective, but certain techniques and methods must be employed by the person who operates the videotape camera. Experienced videographers know how to properly compose shots so that they are aesthetically pleasing. (See the glossary for definitions of the following terms.) Good camerapeople incorporate the photographic principles of framing; they allow for headroom and leadroom; they tilt, pan, dolly, truck, and zoom on cue at correct speeds and with a steady hand. If the subject moves, they follow the subject and they record the shot called for in the script. There are many other aesthetic qualities that comprise a "good" shot. Rely on the experienced cameraperson to capture them.

Visuals must also be technically correct. The light that falls on the subject must be sufficient. In addition, videographers must know what filter to use for different lighting situations, how to manipulate the f-stop for great or shallow depth of field, how to pull focus and a host of other techniques.

Sound

Sound supports the visuals. If a competent scriptwriter has written your script, it will detail exactly what each sound is and when it should occur. Audio tracks may include the following types of sound: spoken word, music, sound effects, ambient or natural sounds, silence and wild sound. Each of these types of sound has a specific purpose and helps communicate your message. (For

148   MEDIA FOR BUSINESS

more detailed information about the different types of audio, refer to Chapter 4, Audio).

Finally, we would be remiss if we failed to warn our readers about a well-known phrase used in the video production business: "Don't worry, we can fix it in post." This seemingly harmless reassurance is an effort to smooth over an error committed during the production process, be it a technical, creative or administrative faux pas. But don't listen. If you record bad audio, re-record it. If an actor mispronounces even an insignificant word, do the scene again. If the camera operator's zoom is unsteady, reshoot the take. In other words, correct any known errors or miscues made in production before you go to post-production. Experience has shown that fixing it in post is much more time-consuming and expensive than making immediate corrections during production.

POST-PRODUCTION

Post-production is the final production phase of a videotape program. During this phase, the program's visual and sound elements are combined in the editing process. Using the script as a guide, the editor and director electronically arrange the scenes in chronological order through the use of a videotape edit controller (see Figure 11.3). The editing equipment allows the editor to select the specific transition device—a cut, dissolve or special effect—called for in the script. One scene progresses to another by way of these transition devices.

The two types of videotape editing are insert and assemble editing. In their book *Small Format Television Production,* Compesi and Sherriffs define assemble editing as "the process of adding new information to a tape shot by shot, or scene by scene, in sequence." Insert editing, the more popular and practical form, is defined as "the process of inserting a shot or sequence into a pre-existing sequence." Insert editing allows greater flexibility and creativity in the editing process. Audio can also be insert- or assemble-edited. Most videotapes accommodate two tracks of audio information. This means you can lay down music, narration, sound effects, natural sounds or silence onto one or both audio tracks. (New video recorders allow four tracks of audio, but as of this writing, two-track units still dominate the market.) If you wish to use more than three sound sources in combination, you must mix them with the aid of an audio mixer and then distribute the sound to your choice of audio tracks. The possibilities are endless if you use a good audio board operated by a competent person.

Figure 11.3: Videotape edit controller

Courtesy of Sony Corp.

Other important activities that take place in the post-production phase include making duplicate copies of the videotape program and labeling and distributing them. These tasks may be completed in-house or at commercial duplication houses in most metropolitan areas.

FORMATS AND EQUIPMENT

Videotape equipment has become the medium of choice in the last decade. Consequently, there are so many choices of equipment and format it takes an expert to decipher all the options. Also, the technology is evolving so fast that by the time you are ready to purchase equipment, a better, less expensive version is on the market. Rather than explore all the choices, we will identify specific formats and types of equipment needed to produce a videotape program. Our discussions in this section will, for the most part, focus on industrial equipment and exclude consumer hardware.

Table 11.1: Comparison of selected tape formats

	COST	PORTABILITY	DUPE QUALITY
3/4-inch	Medium	Fair	Fair
3/4-inch SP	Medium	Fair	Good
1-inch C	High	Poor	Excellent
1/2-inch VHS	Low	Good	Poor
Betacam	Medium	Excellent	Very good
S-VHS	Medium	Good	Fair
Beta SP	High	Excellent	Excellent
M-II	High	Excellent	Excellent
8 mm	Low	Excellent	Fair

The Various Formats

First let's deal with formats. In this instance, format refers to tape size and type. Currently, the formats of choice for nonbroadcast production are 3/4-inch, 1-inch Type C, 1/2-inch VHS and Betamax, and Betacam. S-VHS, Beta SP, M-II and 8mm are also gaining popularity and acceptance. Each of these tape formats has distinct characteristics, costs and uses. Table 11.1 outlines the general characteristics of each format.

Although 3/4-inch tape is in competition with a number of new formats, it remains the industry standard for production. This is partly due to its compatibility. A tape recorded with one brand of 3/4-inch equipment will play back on a different brand or model. This format still produces acceptable visuals and audio, but it loses quality after two generations of duplication.

A new entry in the format field is 3/4-inch SP, a step up in quality and price from regular 3/4-inch tape. It remains to be seen whether traditional users of 3/4-inch tape and equipment will convert to this format.

One-inch Type C is a high-quality format, offering durability and picture and sound quality. It is an excellent format on which to shoot original or "master" footage, but because of its size and bulkiness, the equipment is rarely taken to remote locations. Instead, it is principally used in studio settings. It is also the best format for editing, particularly when special effects are used. Its main drawback is price; it is by far the most expensive format to use in nonbroadcast applications. Prices for 1-inch tape stock range from $1.75 to $2.50 per minute.

Industrial 1/2-inch formats (VHS and, to a lesser degree, Beta) have gained widespread acceptance. This type of videotape is best for play back because so many companies and agencies own 1/2-inch recorders and players. Production equipment, however, does not yet approach the standards of 3/4-inch.

Betacam® production equipment and tape provide a great deal of versatility. Betacam® equipment is extremely portable because the camera includes a built-in compartment for recording the videotape. No longer does the cameraperson or a production assistant have to lug around a recording deck. This makes Betacam® a favorite of commercial and nonbroadcast producers.

One relatively new entry into the videotape field is S-VHS. This tape can be played back on standard VHS players but it produces a better quality image when recorded with and played on S-VHS equipment. Its duplication quality approaches that of 3/4-inch tape. Whether this tape and its associated equipment will have an impact on the nonbroadcast market remains to be seen. Some analysts predict that it will replace 3/4-inch as the nonbroadcast standard.

Beta SP® and M-II® are also relatively new competitors in the field. Both are 1/2-inch formats that produce remarkable picture clarity and sound quality. In addition, they can be duplicated up to six generations without a noticeable loss in picture quality. Cost can be prohibitive for those with small budgets.

Sony introduced 8mm with the intention of offering an alternative to the 1/2-inch consumer tapes (VHS and Beta). Although the equipment is lightweight and very portable, picture quality is comparable only to that of VHS. It should therefore not be used as a master tape format to be edited.

Over the next few years, many of these formats will become obsolete as the digital recording revolution proceeds. Ampex and Sony have introduced digital recorders capable of recording and duplicating programs with no loss in picture or sound quality. This is the future of videotape recording, but at present, cost remains a major obstacle. It won't be until manufacturers reduce prices that digital recorders will take over as the industry standard.

The Equipment

Cameras, recorders, microphones, tripods, lights, teleprompters... more than any other visual medium, videotape production involves a plethora of equipment. And, unfortunately, every conceivable letter and number is used to describe different models. What follows is a brief description of some of the video equipment used to produce a videotape program.

Cameras

Video cameras perform one major task. In simple terms, they convert light and images to an electronic signal which is recorded onto tape. Currently, most nonbroadcast producers use "tube" cameras which contain one or three individual pickup tubes that correspond to red, blue and green. The trend, however, is toward CCD (charged coupled devices) cameras which have a greater sensitivity to high-contrast images than tube cameras. Whereas tube cameras may be damaged by intense light, CCD cameras are unaffected. As they are improved and become less expensive, CCD cameras will undoubtedly replace tube cameras.

At first glance, video cameras may appear confusing. They have all sorts of buttons and switches and come equipped with a variety of options. In the following section, we will discuss the basic components and controls found on most cameras.

The majority of videographers use a zoom lens with their video cameras. The zoom lens allows the camerperson to bring a subject closer or make it appear further away simply by changing the focal length of the lens, either manually or automatically. The white balance control plays an important role in determining correct color. To set it, fill the viewfinder with a pure white object (paper, cardboard, etc.), then depress the white balance control. This ensures proper color during taping so that faces, for instance, don't appear green or yellow. This procedure must be repeated every time the camera is turned on.

Many cameras feature a pre-set white button which balances the camera for either daylight or tungsten light conditions. Some cameras include an automatic gain control (AGC), an iris control that automatically adjusts the lens opening based on the amount of light. If the light level is too low, the AGC automatically opens up the lens to allow more light to enter. Conversely, if the light level is too high, it closes the lens down, thus reducing the amount of light. This function is both a blessing and a curse. In situations where the light level constantly fluctuates, it has great usefulness. However, it can pose problems. Suppose you are taping a person in a hospital corridor. As the person delivers the on-camera narration, some doctors and nurses in white coats walk into the picture. The camera iris will adjust to the white coats and close down, allowing less light to enter the camera lens. When this happens, the image of the on-camera talent appears dimmer and more difficult to see.

Video cameras can feature a host of "bells and whistles." The trick is to know your camera's capabilities and make use of them as best you can. Don't pay for options you don't need.

Videotape Production 153

Recorders

Another necessary piece of equipment is the videotape recorder (see Figure 11.4), which records the video and audio signals onto tape. AC power is usually used to operate the recorder and camera system, but when power is not accessible, batteries can provide the necessary current. Since most recorders are separate from the camera, they have to be carried by the cameraperson or an assistant. The notable exception to this is the Sony Betacam® which has a separate compartment in the camera where videotape is placed for recording.

Lights

Many good shots have been ruined because a videographer failed to use lights. Without adequate lighting, the video camera is not able to record the image. Many videographers and their crews use a portable light kit which contains a minimum of three lights with stands. With these three lights, subjects may be illuminated through triangular or three-point lighting (see Figure 11.5). This serves as the foundation for many types of videotape lighting.

Figure 11.4: Videotape field recorder

Figure 11.5: 3-point or triangular lighting

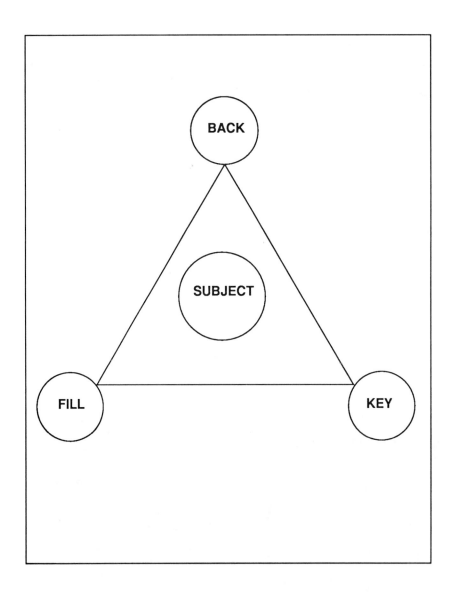

Microphones

Microphones are most often used to record voice and natural sounds. They vary in size, shape and pickup pattern. In the past decade, their portability and quality have dramatically improved. Microphones have become smaller and more versatile and possess greater sensitivity and selectivity. This has led to greater emphasis on the sound aspect of videotape recording. Remote recording has become simpler with the advent of the wireless microphone (see Figure 11.6) because talent has more mobility without the worry of getting tangled up in a microphone cord. Directors can shoot from farther away now that the audio person no longer has to run hundreds of feet of cable. Although recording sound is easier than ever to accomplish, it is still an art. Electrical interference, unwanted background sounds, and improper volume levels still plague the audio portion of some videotapes. Our advice is to know your equipment or hire a competent audio technician who does.

Figure 11.6: A wireless, lavalier microphone

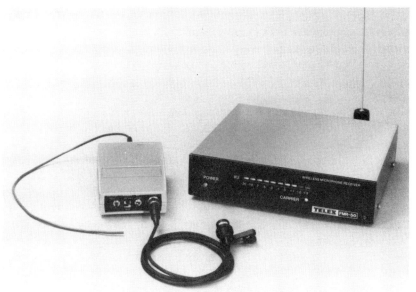

Courtesy of Telex Corp.

Other Equipment

Peripheral equipment associated with videotape production includes tripods, monitors and teleprompters. A cameraperson should use a tripod (see Figure 11.7) whenever possible. The head of the tripod, on which the camera rests, allows the camera to execute zooms, pans and tilts. The smoothness and steadiness with which the camera moves depends on the type of head. There are two types of heads, fluid and friction. A fluid head affords smoother, steadier camera movement, but it is much more expensive than a friction head. Other tripod accessories facilitate various other camera movements. When placed on wheels, also called dolly or pedestal assemblies, the cameraperson can smoothly move the tripod for dolly, truck and arc shots (see Figure 11.8).

Monitors, or television sets that do not receive broadcast stations, are essential for both studio and field recordings. In the studio, the director needs them to see each camera shot, to preview special effects and ensuing shots, and to see what is actually being recorded or transmitted. In the field, monitors are valuable for setting up camera angles, observing the action while it is videotaped and playing back each recorded scene to verify correctness.

A teleprompter (see Figure 11.9) is another accessory used in both studio and field settings. This device fits onto a video camera and scrolls the written script. It permits the on-camera person to look in the lens and read the script aloud. A teleprompter is especially valuable for those who have trouble remembering their lines or if you do not have time for multiple takes.

Other video equipment is used almost exclusively in the post-production phase. This includes videotape editing, audio and graphics components. Figure 11.10 illustrates a conventional videotape editing system which would include an edit controller, a videotape player (called the slave) and a videotape recorder (called a master). Editing machines transfer audio and video information to a master tape in the proper sequence as called for in the script. As we discussed earlier in this chapter, editing machines perform either insert or assemble edits. Like all videotape equipment, editing systems vary in price and capability. They can cost as little as $3000 or as much as $300,000. In most cases, the adage "you get what you pay for" proves true. The lower-end editors are capable of "cuts only" editing. This means that they cut from scene to scene. When they are linked to switchers and other specialized equipment, the more expensive systems perform cuts, dissolves and special effects from one scene to the next.

Audio mixers are also found in videotape editing facilities. When multiple audio sources are used, mixers combine the different sound elements onto

Videotape Production 157

Figure 11.7: ENG fluid head tripod

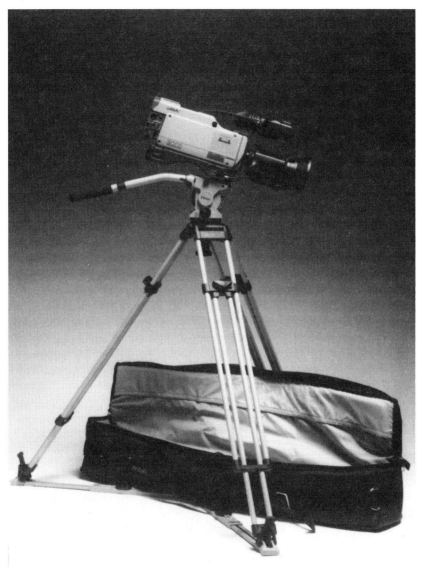

Courtesy of Vinten Equipment Inc.

Figure 11.8: Dolly, truck and arc camera movements

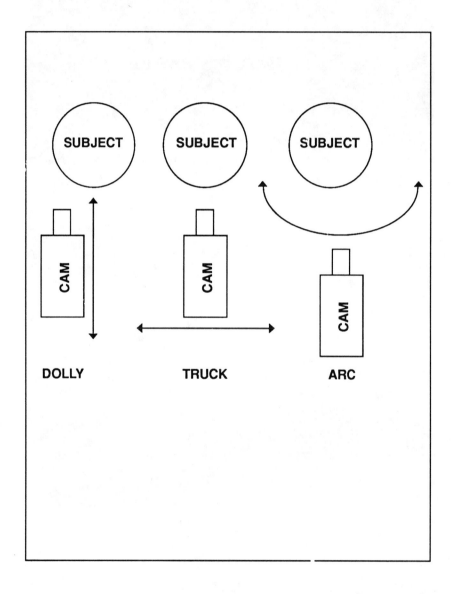

Videotape Production 159

Figure 11.9: Teleprompter (front) mounted on a video camera

Figure 11.10: Videotape edit system

160   MEDIA FOR BUSINESS

one or both of the audio tracks of the videotape. Without an audio mixer, many videotape productions would be limited to two sound sources. Audio mixers simplify and condense complex soundtracks so that they can be effectively used in videotape programs.

Graphics are a common form of visual information in videotape productions and are usually produced by electronic devices called character generators. As Jim Sullivan, president of PenComm Productions in Denver says, "A character generator is like an electronic video typewriter." The simplest character generators (see Figure 11.11) superimpose black or white letters over videotape. More sophisticated models produce a variety of letter sizes and font styles and a limited number of symbols and color backgrounds over which you can superimpose graphic information. More complex graphic systems (paint boxes and animators) allow the operator to digitize information, render freehand drawings and carry out animated movement. These systems require extensive operator training and large client budgets if they are to be cost effective.

Figure 11.11: A character generator

SUMMARY

Videotape constitutes the fastest growing medium in the field of visual communication. Videotape players are widely accessible and easy to operate and the programs are very portable. As technical advances continue, videotape equipment will become cheaper, smaller and more compact, while the quality improves. Yet, as is the case with every medium, equipment alone will not guarantee a successful program. Program content and production techniques will still determine a program's effectiveness.

12

Interactive Video

Let's begin with another true-to-life dialog:

Jessica: Hi, Paul. Glad I ran into you. Our department needs a program for a new product line. Maybe you could help me.

Paul: Sure. What do you need?

Jessica: Well, we need a training piece for our field representatives. They need to know how to operate the new equipment. They all have different learning levels, so the program needs to fit everyone's needs. Think you could help us out?

Paul: As a matter of fact, this type of program is tailor-made for a technology we're just getting involved with—interactive video.

Jessica: Interactive? You mean when someone accompanies the program and talks to the visuals.

Paul: No, not really. As a matter of fact, if produced properly, this program won't require an instructor for the training. Would you like to see an interactive video we just completed?

Jessica: That sounds interesting, but I've got to warn you, I'm skeptical about not having a trainer present.

Paul: That's all right. I think you'll change your mind after you see the program.

164   MEDIA FOR BUSINESS

Interactive video programs have been in existence for a number of years, but many people like Jessica are still unaware of their capabilities, the differences between interactive and standard video programs and their training applications.

WEIGHING THE DIFFERENCES

In this chapter, we'll bring you up to date on these points and provide a step-by-step production guide. But first, what do we mean by interactive video? Interactive video, a spin-off from more traditional videotape production, requires active rather than passive involvement from the viewer; it is designed with viewer participation in mind. Interactive video programs differ from other video programs in the following basic ways: 1) a knowledgeable training specialist or an instructional designer is essential for their development and production; 2) videotape or film production constitutes just one phase of their production process, so the roles of producer and director are diminished; and 3) production equipment must produce high-quality video and audio signals. Additional, less basic differences will be discussed later in this chapter.

A word of advice before we continue: very few in-house media centers have the expertise necessary to produce an interactive program. If you need to develop one, consult a specialist or the nearest International Interactive Communications Society chapter in your area for information. These experts are not as numerous as those who have video production backgrounds. However, those with experience will be able to design and produce your programs from concept to completion.

Advantages of Interactive Video

Interactive video programs offer several distinct advantages over other media programs:

- They serve as valuable educational and training tools.
- For the most part, the programs are self-paced.
- Individual segments or entire programs are easily repeated.
- They can test a viewer's comprehension of subject material.
- They provide a consistent message.
- They can store immense quantities of information if produced on videodisc.

- They save costs by eliminating the need for individual instruction or training.

Let's look at these in more detail. Interactive programs are particularly effective in educational and training settings because they can present a large body of information while allowing each viewer to learn at his or her own pace. Important information may be repeated until the viewer understands it while the advanced viewer can skip ahead. Such self-pacing capacities are unique to interactive video programs.

Testing is another important option interactive programs provide. Pre-testing allows the viewer to proceed to the part of the program which best suits his or her knowledge level. During the program, built-in components at the end of chapters or units test individual comprehension rates. The post-test at the end of the program not only measures the viewer's comprehension, but also suggests remediation strategies based upon the viewer's test score. This helps trainers assess viewer abilities and determine individual progress. Most interactive systems can easily score and tabulate test results, which can be printed out to hard copy and kept by the viewer.

By providing information in a consistent way, interactive video programs can take the onus off trainers when they are "having a bad day." Delivery of information is no longer dependent on the individual presenter's idiosyncrasies. The trainees receive the information in a tried and tested way—one that will make the learners receptive to the message, not one that will make them reluctant to learn.

Videodisc storage capability is far greater than that of standard videotape. Although the running time of most interactive videodiscs is limited to 30 minutes, its ability to store up to 54,000 still frame images far surpasses most videotape storage capabilities. (We will discuss videodisc later in this chapter.)

Although some trainers view interactive video as a threat to their jobs, the technology actually affords trainers and interactive program planners the opportunity to revolutionize the training field. Interactive programs combine new technology with individualized learning and greatly affect the way people learn. Interactive video does not make trainers obsolete; it simply provides them with more advanced tools that enhance the entire training process. Instead of spending an inordinate amount of time presenting the training information, training departments can focus on identifying and solving problems for their companies. This is a major benefit which any open-minded trainer must recognize as an aid to company productivity.

There are other advantages. By eliminating the need for a "warm body" to present training information, companies save personnel costs. Trainers no

166   MEDIA FOR BUSINESS

longer have to accompany and deliver training packages. Travel, lodging and other associated expenses are greatly reduced because training departments can mail a videodisc to perform the same function as the trainer.

There is one caveat to this cost-saving rationale: the subject matter covered in an interactive program should be of long-term relevance to a company. If it will become rapidly outdated, it will not be worth the investment required for production.

Disadvantages of Interactive Video

Like any other medium, interactive video has disadvantages. These include:

- longer production times than for most other media
- higher production costs than for most other media
- difficulties associated with mastering and duplication
- playback equipment that can be costly and complicated to operate

Because the production of an interactive videotape often involves more personnel, equipment, resources and planning than other media, it stands to reason that production time will be longer than that of the more traditional media we have already covered in this book.

Increased production time is one contributing factor to greater production costs. Without relating specific budget line items, it will become obvious that interactive programs can be expensive. However, even though production costs are greater than those for most other media, it is necessary to weigh these costs against program goals, objectives and outcomes.

In any business, time plus personnel equals money. That's true for media production and it's especially true for the production of interactive programs. Add to these factors the need for specialty equipment (both hardware and software), and costs continue to rise. Added planning, computer programming, testing and evaluating phases also require funds. Later in this chapter, we'll show you how these relatively high production expenses are offset by reduced educational and training costs.

Interactive programs produced for videodisc require additional post-production steps—the mastering and duplication processes. "The final few steps of the process can easily take from eight to ten weeks to complete, and each step is critical," Richard Schwier points out in his book *Interactive Video*. Since the number of videodisc mastering houses in the U.S. is limited (five at this writing), producers sometimes encounter delays in the mastering and replication of their discs. For this reason, we advise scheduling the final phase of post-production in advance.

Interactive programs can be played on a variety of systems. To the novice, a typical interactive playback system may look imposing and intimidating. Some systems use an external computer or a player that has internal memory capability. Others must have software loaded into them before they become operational. Once in operation, most interactive programs require feedback from the viewer. This may involve touching a screen or typing on a keyboard. Regardless of the system, each one has functions that are outside the realm of traditional videotape equipment. Since the operation of the interactive equipment requires more than simply inserting a tape in a machine and viewing it, this process may confuse the user and interfere with the educational or training process.

Equipment cost poses a further problem. Interactive playback machines and their associated equipment are more expensive than traditional videotape players. In order to play Level III programs (see our discussion of formats in the following pages), not only must you purchase video monitors and videodisc or videotape players, but you must also buy the necessary computer equipment—CPU, keyboard, monitor and sometimes software. To complicate matters, individual systems are not normally compatible with one another. This means that commercially produced interactive programs that may fit your needs may not match your hardware and software.

INTERACTIVE VIDEO FORMATS

Interactive programs may be produced on videodisc or videotape. The videodisc is "a slim, rapidly spinning circle of plastic that brings sound and color to a TV screen."[1] It is much like an audio compact disc, except that it is about the size of a record album.

The most commonly used videodisc utilizes the optical-reflective system. "This system supports interactive applications, provides impressive quality, durability, and capacity, and exploits training, industrial, and commercial opportunities."[2] Most of these laser videodiscs contain 54,000 separate video

---

[1]Sigel, Efrem et. al. *Video Discs—The Technology, the Applications and the Future* (New York: Knowledge Industry Publications, Inc., 1980).

[2]Schwier, Richard, *Interactive Video* (New Jersey: Educational Technology Publications, Inc., 1986).

frames on each side of the disc. Each frame is capable of storing one picture with far greater clarity than does videotape. When inserted in a videodisc player, a laser light reads the information from the disc and then transforms it into video signals. The disc also contains two audio tracks which are read and transferred as audio information. Because the laser does not come into physical contact with the disc, the videodisc maintains its integrity indefinitely.

Videodiscs provide another critical capability that videotape cannot offer: random access. This allows the viewer to go to any part of the program within a three- to five-second time span. The viewer does not have to watch preceding or following segments to reach the desired point. Currently, most videodiscs are read-only; very few can record information (see Figure 12.1). A few companies, however, have introduced videodisc machines that can record information directly onto the disc.

Interactive video programs may also be produced and played on standard videotape. However, all interactive videotapes are linear. In other words, the viewer must access other parts of the program by viewing the entire program, rewinding, or fast forwarding, the tape through various sections of the program. This takes time and reduces the viewer's attention span. In addition, videotape

Figure 12.1: Laser videodisc player

deteriorates each time it is played. After being played repeatedly, the interactive videotape program will have to be replaced.

Levels of Control

Presently, there are three or four levels of control afforded by interactive videodisc programs. As of this writing, Level IV does not have full acceptance or concurrence within the interactive community. For this reason, we will limit our discussion to the other three levels.

Level I consists of a system in which the viewer presses buttons on a keypad that comes with the videodisc. In most instances, the controller simply moves the program forward or backward. Level II systems offer greater options. A Level II videodisc has a memory and can store information. It has the capability to accept loading of a short computer program that controls the videodisc player's actions. In some programs, users may input information into the program. In a Level III system, an external computer controls the videodisc and accepts input from the user. The computer has a color monitor that can display both video and computer-generated images. Generally, the user can interact with the program to a greater degree with a Level III system. Level III systems also offer touch screens and printout devices that are configured into the system.

Budget Considerations

As we discussed earlier in this chapter, budgets for interactive video programs are typically greater than those for other media. Costs are higher because interactive programs involve more than just writing, shooting and editing a videotape. They usually require additional personnel. Individual team members may be needed to plan and carry out each of the processes. For this reason, the development team (personnel) selection is a major budget consideration.

Personnel

An interactive video team includes all the personnel necessary for any type of major production—a scriptwriter, producer, director and production crew, plus the following:

- project manager
- subject matter expert
- instructional designer
- computer programmer

The project manager oversees each aspect of the project and must have experience in instructional design or training as well as the interactive video implementation process.

The subject matter expert needs to have an intimate working knowledge of the topic and will be called upon to provide facts and statistics and to proof what the writer has scripted. In addition, this person must check the flowcharting process to assure that all the options have been considered and that all options are logical.

The instructional designer works closely with the project manager and subject matter expert. At times this person may also assume the role of project coordinator. Typically, an instructional designer establishes and outlines the learning/training strategy. This person ensures that the project goals are being met, the program learning objectives are measurable, and that the audience will somehow demonstrate what it has learned.

The computer programmer works with an authoring tool (a programming language) and determines the menu selections and commands which enable the program to operate properly.

Other persons, including specialists who understand the flowcharting process, additional content experts and other technicians may be necessary, depending on the project.

Occasionally, some of the development team roles may overlap, but interactive video programs must be fully and properly staffed if they are to succeed. Our recommendation is simple: hire qualified people and don't overwork them.

Equipment

Purchase and/or rental of production equipment may balloon the budget, especially since high-quality video equipment should be used. When recording onto videotape, it is advisable but not mandatory, to master on 1-inch equipment. Its superior signal will produce better field footage which, in turn, results in higher-quality master and duplicated discs. In some cases, motion picture film (16mm or 35mm) is preferred, although it will increase production costs.

Special services, including complicated graphics, special effects capabilities and compressed audio recording, may come into play during the production process. All these production tools and services need to be considered in the budget.

If the program is to be distributed on videodisc, mastering and replication are additional post-production processes to factor into the budget. Although these costs can be significant, large quantities of duplicates are comparable in cost to videotape copies.

In addition to the video recording equipment, you will need videodisc players and computer equipment for play back of the program. As you can see, all of these equipment costs translate to greater production expenses.

PRODUCTION DESIGN CONSIDERATIONS

Although interactive programs incorporate the same scripting and production techniques used to produce a videotape, they are different in many ways. These differences are best demonstrated in a listing of the implementation process as provided by Stephen Floyd in *The Handbook of Interactive Video* (1984). According to Floyd, there are 10 distinct steps for implementing an interactive program:

1. front-end analysis
2. instructional or evaluation strategy
3. flowcharting
4. scripting
5. production
6. assembly and programming
7. pilot evaluation
8. revision
9. duplication and distribution
10. project evaluation

The purpose of the front-end analysis is to define the problem to be resolved in measurable terms and to establish training objectives.

Instructional or evaluation strategies are used to determine how and when to evaluate subject proficiency and to define acceptable competence levels (by test scores, etc.). During this stage, the content expert identifies potential trouble spots in the program.

A flowchart (see Figure 12.2) is like an organizational chart. It shows the logical progression of the program and how different elements are interrelated. In this phase, all the possible options the program might contain are diagrammed. As Nicholas Tuppa writes in his book, *A Practical Guide to Interactive Design* (1984), "Flowcharts are used for a number of different reasons. They are used to communicate general program flow to clients and writers, to indicate the organization of all parts of the program to production staff, and to indicate the mechanics of the program to computer programmers."

Flowcharts are related to branching, which provides the viewer with menu options from which to choose. Branching is a characteristic which allows the program to move from one point to another. Figure 12.3 shows a simple example of branching, the kind that would be found in a Level I interactive program you might encounter in a shopping mall.

The scripting and production phases of interactive are much the same as for regular videotape production (refer to Chapters 3 and 11). However, writers, producers and directors must work closely with the project coordinator because there are more variables to consider in the interactive production process. For example, the writer has less latitude in determining specific content and must heed the flowcharting options and focus on creative ways to present that information. The same may be true for the producer and director. Instructional strategies and flowcharting may determine the shooting style to a great extent.

Don't forget that technical and engineering standards must be adhered to in the production process in order to maintain the quality of the final product— the master disc. For this reason, we strongly advise having a professional engineer assist in the production and editing processes. This person will help ensure that your program meets all the technical requirements.

After production, the assembly and programming of the software and its hardware system begin. At this stage, an authoring tool is chosen. Authoring tools are divided into two major categories: authoring systems and authoring languages. An authoring system is easier to use because it does not require mastery of computer programming. The authoring language allows greater flexibility, but requires an understanding of computer programming. A programming language provides command or menu selections which guide the direction of the program.

This assembly and programming phase also represents the last opportunity for the project coordinator to select a distribution format (tape or videodisc). After software and hardware selections are made, a specialized computer programmer develops and inputs commands that enable the computer and its software to properly interact with the videodisc player.

Interactive Video 173

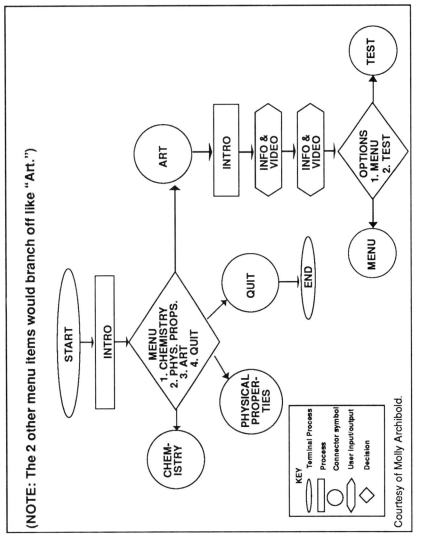

Figure 12.2: Sample flowchart—"The Art and Science of Making Bread".

(NOTE: The 2 other menu items would branch off like "Art.")

Courtesy of Molly Archibold.

174 MEDIA FOR BUSINESS

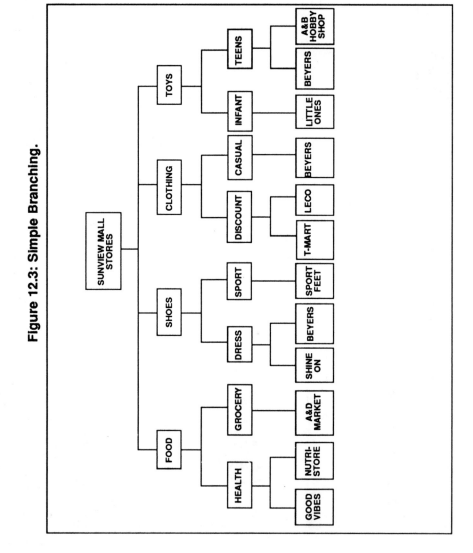

Figure 12.3: Simple Branching.

Pilot evaluation, testing and debugging of the program come next. During this stage, the program is tested for technical quality, equipment compatibility and program content. This often involves asking a "typical user" to run the program and operate the equipment as prompted by the computer. This stage helps the program designer and content expert discover design flaws in the equipment or flowcharting process before the program is duplicated, distributed and sent out for use.

After the content expert has approved and incorporated suggestions from the evaluation process, the program is ready for duplication and distribution. The duplication process varies, but most videodiscs can be duplicated within a six- to 12-week period.

The last step, project evaluation, is just as important for interactive projects as for any other media production. It provides an ongoing analysis which makes production of subsequent programs easier and more cost-effective. Project evaluation also points out what aspect of the program design was successful and what should be avoided in the future.

Considering that interactive programs usually require longer production schedules, more personnel, additional equipment and bigger budgets, why would anyone want to produce an interactive program? The answer is ironic: cost savings. Properly used, interactive video is less expensive than other forms of training.

APPLICATIONS

Interactive video programs have many applications including the following:

- education
- training
- marketing
- archival development
- entertainment

The value of interactive programs in education cannot be overestimated. They accommodate self-paced learning, provide individual attention, present information in a consistent manner and serve as an excellent time management tool. In his 1987 book *Videodisc Training: A Cost Analysis,* Richard Brandt identified another reason for using these programs: "Interactive video may be applied when teachers with subject matter expertise are in short supply." Program content ranges from the teaching of foreign languages and customs to topics such as health, data processing, management and industrial technology. Currently, there are nearly 450 off-the-shelf courses available to users of IBM systems.

Interactive programs are especially attractive to trainers for many of the same reasons educators like them. According to Brandt, "Disc-based training should be considered when there are a large number of learners distributed over time and place." He also concludes that "Videodisc systems are particularly suitable when training requires continuous practice and/or retraining."

The use of interactive training programs in the manufacturing industries may be especially valuable. These programs are produced to help train people to perform specific tasks such as machine operation, assembly line production or routine maintenance. Interactive programs also allow for simulation of situations that could not otherwise be experienced. For instance, a program may demonstrate how to properly dispose of chemical waste without endangering the learner. This allows trainees to vicariously experience the situation and become familiar with the correct procedures in a safe manner. In addition, procedures that require steps to be followed in a specific order—CPR, accounting, cooking, etc.—can be easily taught.

Interactive programs can also present sales and management techniques. Training methods are applied to a variety of topics ranging from demonstrations on how to conduct telephone sales and departmental meetings to product identification and banking procedures. Brandt points out that "they may be useful tools in training that involves such skills as problem solving and decision making." Interpersonal relations offer another training application for interactive video. For instance, we know of at least one program that demonstrates proper employment interview techniques and illustrates how to "size up" job applicants.

For marketing purposes, interactive videodiscs are useful as point-of-purchase programs. Whether the disc is mailed to a potential client or viewed on the exhibit floor of a trade show, the program serves as a sales tool. It may stand alone or augment a sales presentation. Other interactive point-of-purchase programs show promise in the restaurant, retail and tourist industries. Regardless of its use, the interactive video program is gaining popularity as a new and efficient marketing tool.

Since videodiscs are capable of storing immense quantities of information, they are also valuable as archival tools. Libraries use them to store catalog and research information; hospitals use them to store medical photos that assist in teaching, as well as operations procedures and research. Companies that formerly used print catalogs to display their inventory are turning to videodiscs as a way of presenting their line of products. Such companies find that products shown on a videodisc look more realistic and can be presented in their appropriate environments. Also, videodiscs are lighter than print catalogs and can reduce postage expenses. After customers place an order, the videodisc can be stored for future use or returned and recirculated to another client.

Videodiscs also offer an alternative in the entertainment field. Although 1/2-inch VHS is the most common format that consumers rent to view motion pictures at home, hundreds of theatrical films are available on videodisc. Although the programs do not fit the definition of "interactive," they offer superior visual and audio quality, thus enhancing the home viewing experience.

Interactive games provide another form of entertainment. In his book *A Practical Guide to Interactive Design,* Nicholas V. Iuppa states, "Some games can take a viewer through long processes as complex as an entire economic cycle or military campaign. Others focus on a few special skills such as throwing a ball through a hoop." Action games are popular in video arcades and as programs to be played on home computers. They involve some sort of decision-making process. Adventure games rely more upon the skill of the individual player.

SUMMARY

Although interactive video programs have not become as popular as initially predicted, they continue to gain prominence in the training and education fields. They are unmatched by any medium when self-paced, individualized instruction and training are paramount. They are also valuable when a consistent message must be delivered over time to a disparate audience. In this regard, interactive video programs are more cost efficient than other types of training, motivational, sales and instructional media programs.

Although interactive video involves a complex development process that requires greater production time than other types of media production, the high cost of production can usually be justified when specific learning situations exist. If, however, certain training conditions, learning environments or subject matter criteria do not exist, then the decision to use interactive video may be inappropriate and costly.

13
Teleconferencing*

INTRODUCTION

"Well, we could have a teleconference," she said, uncertainly. In her mind she saw all the branch office personnel huddled in their conference rooms, paging through charts while interacting with a speakerphone on the conference table.

"You must be crazy," the other woman said. "Where would we get a satellite? And why would we want to see their faces?"

Part of the confusion about the role of teleconferencing in business is that no one is sure what we mean when we say "teleconference." The International Teleconferencing Association (ITCA) Program Committee indicates that "teleconference" is "an industry term for . . . audio, audiographic and video (conferencing)" which have in common "the use of a telecommunications channel and station equipment linking multiple locations for interactive communications that involve the active participation of users."

"Teleconference," then, is an umbrella term that is useful for generic reference to synchronous communications between or among groups in different locations. (I say "synchronous" deliberately to exclude electronic bulletin boards, electronic mail and what in the past has been called the "computer conference.")

Having narrowed the scope somewhat, let's look at four subsets of the teleconference.

---

*This chapter is adapted from portions of *The Teleconferencing Manager's Guide*, edited by Kathleen J. Hansell, published by Knowledge Industry Publications, Inc., White Plains, NY, 1989.

For communications where the spoken word, perhaps augmented by paper charts, is sufficient to accomplish the business at hand, the *audio conference* provides an easily established form of teleconference capable of involving people in even the remotest of outposts. Using loudspeaker telephones and bridging systems, the audioconference is one of the least limited of the teleconference technologies. It is also the basis for three additional forms of teleconference.

When the exchange of spoken words is insufficient, an audioconference can be enhanced with electronic graphics components that can display word charts, diagrams, text materials and still video images of people or objects. With the increased use of personal computers in business, the *audiographic conference* has found growing acceptance, especially for training and educational purposes, where heavy visual interaction, including the real-time annotation of images among all sites, is necessary to accomplish the communications objective.

For group communications requiring a more intimate and active exchange of visual information among the participants, the interactive *videoconference* adds motion video to the audioconference. Of all the teleconference technologies, the videoconference most closely approximates an in-person meeting. Primarily a point-to-point technology at the moment, the multiple-site videoconference, where all groups can see and hear each other, is the next wave.

For situations that require visual communication from a source to multiple locations, *business television* adds live broadcast television to the audioconference. Whether transmitted on permanent or temporary networks, business television has the potential for high-impact, fast-paced visual communications that reach widely dispersed audiences.

Table 13.1 shows the commonalities and differences among these four forms of the teleconference. While I have compartmentalized every aspect of these teleconferences for the sake of discussion, in the real world nothing is quite so simple. This is especially true in the realm of communications, where there is considerable overlap. And while it is useful to differentiate the technologies in terms of their uses, their hardware and their benefits, to answer the question, "What is teleconferencing?," it is also useful to look at the bigger picture and explore "What can it do for me?"

For most people who ask the "What can it do for me?" question, the first thing that comes to mind is the possibility of avoiding a trip to a meeting or training session. For some, this is inviting; for others, disappointing.

In the 1970s there was a great deal of research on the "travel/transportation tradeoff" of telecommunications. The premise was simple enough: Instead of moving bodies, let's move information. To understand the capacity for doing this, travel studies were widely conducted in business and government organizations. Some of the numbers were pretty impressive.

At the same time, there was a growing body of evidence that telecommunications didn't displace travel in direct proportion to its use. In fact, in some instances travel increased with the introduction of telecommunications.

Not so astounding, I thought. After all, Alexander Graham Bell's first use of the telephone was to invite poor Watson to physically move from his location to Bell's. And the telephone never did negate the need or desire for business or personal travel. Why should teleconferencing be any different?

Appreciation for the softer benefits of teleconferencing was already evident, e.g., "quicker decisions," "less information float," "access to experts, etc. What is only recently coming to the forefront, now that we have some experience under our belts, is that the real value of teleconferencing systems is not in their ability to *save* money or *avoid* travel. Rather, the teleconference gives companies the strategic capability to position their organizations and their products in the marketplace. Boeing gets its 747 to market months sooner; Merrill Lynch reassures customers in the wake of the stock market crash; Eastman Kodak gives students an opportunity to interact with master photographers. These are the real benefits. And they have little to do with saving money or avoiding travel.

Since time immemorial, humans have always chosen the appropriate tool for the task. Teleconferencing technologies give us one more set of tools.

These tools are changing rapidly. The year 1989 marked the 25th anniversary of the Picturephone, introduced at the 1964 New York World's Fair as the communication tool of the future. While we don't have motion video communications in every home, we've come a long way since 1964. Most of the progress has occurred in the last few years. In 1980, when we plotted to introduce digital videoconferencing as a substitution for travel, we were looking at 6 million bits per second as "incredibly" low compression. Today, transmission at 56,000 bits per second can be "dialed up" as needed in a growing number of locations. Business television then required large, unwieldy earth stations; today, small antennas sprout from every office complex. The old "push-to-talk" microphones have been replaced today with microphones so unobtrusive that casual observers don't see them. Still-video images that took minutes to painstakingly paint on screens now appear in seconds on audiographic systems. The strides made in recent years in teleconference technology, and in our understanding of how to use the technology, have been large ones; and the steps we will make in the next few years promise to be even greater.

TABLE 13.1: A MATRIX OF TELECONFERENCING TECHNOLOGIES

	AUDIOCONFERENCING	AUDIOGRAPHIC CONFERENCING	VIDEOCONFERENCING	BUSINESS TELEVISION
CONFIGURATION	Point-to-point or multipoint, multiple number of sites	Point-to-point or multipoint, two or several sites	Point-to-point or multipoint, two or several sites	Point-to-multipoint with multiple receive sites
PRIMARY USES	Business meetings requiring verbal exchanges	Business meetings requiring verbal interaction and the exchange of documents, drawings, and charts	Business meetings requiring motion images of participants or activities	Training sessions; informational or motivational events
VIDEO CAPABILITY	None	Still video, one direction at a time	Motion video, usually two directions at any given time	Motion video, one directio from origination to receive sites
AUDIO CAPABILITY	Fully interactive	Fully interactive	Fully interactive	One way from origination to receive sites; optional return audio via phone lines
MOST CLOSELY RESEMBLES	Telephone conversation	Something between a telephone conversation and an in-person business meeting	An in-person business meeting	A television show
TRANSMISSION	Analog, usually terrestrial	Analog or digital, usually terrestrial for domestic connections	Usually digital, via terrestrial or satellite channels	Analog, almost always via satellite

TABLE 13.1: A MATRIX OF TELECONFERENCING TECHNOLOGIES (Cont.)

	AUDIOCONFERENCING	AUDIOGRAPHIC CONFERENCING	VIDEOCONFERENCING	BUSINESS TELEVISION
COST	Equipment, $100 to $7,000 per site; transmission, standard telephone connection rates; bridging services, $20 per site per hour	Equipment, $3,000 to $50,000 per site; transmission, standard telephone connection rates or digital data rates, bridging services additional	Conference systems, $7,500 to $300,000; codecs, $10,000 to $75,000; transmission, $25 to $1,000 per hour	Receive site equipment, $5,000 to $10,000; satellite transmission, $600 and $2,500 per hour for U.S. and European/international systems, respectively; $300 per hour for fixed uplinking services; production costs vary
SPECIAL STRENGTHS	Readily available equipment and transmission facilities; least expensive of the teleconferencing technologies	Readily available transmission; excellent merging of video and computer technologies	Ability to interact visually in real time; capability for videotapes and other typical meeting media during conference; growing number of public and shared facilities	Ability to transmit uniform messages simultaneously to large audiences; low cost per person reached
LIMITING FACTORS	Inability to share visual images during a conference	Equipment incompatibility	Equipment incompatibility	Special resources required for program production
TYPICAL USERS	All business enterprises	Engineering, architecture, design, creative services, medicine, education	Aerospace industry, defense contractors; manufacturing (pharmaceutical, computer and automotive industries; other high technology enterprises); energy; banking and insurance	Financial services; retailing, manufacturing (automotive and computer industries, high technology enterprises); education and training services

Source: Copyright © 1990, KJH Communications.

184   MEDIA FOR BUSINESS

As a matter of fact, the scene at the beginning of this chapter is probably out of date already, so let's replay it:

"Well, we could have a teleconference," she said, matter-of-factly. In her mind she saw a number of possibilities for connecting the branch office personnel.

"You must be a mind-reader," the other woman said, "I'll arrange it right away."

AUDIOCONFERENCING

Introduction

The most widely used of teleconferencing technologies is the audioconference. It is the most transparent, easiest and most used of the teleconference technologies. It's so elegantly simple that, in many cases, audioconferencing is often taken for granted and not given its due as one of the most significant of the teleconferencing technologies. In fact, it is the basis of all other forms of teleconferencing. Thousands of telephone meetings are interconnected in the United States every day by phone company conference operators, commercial bridging services and private audio bridges.

Audioconferencing is defined as three or more persons at two or more locations conducting oral communication in real time. The interconnection medium is usually the telephone, which can incorporate microwave, satellite, fiber optic or coaxial cable transmission. Companies offering interconnection services have found that the average audioconference links seven or eight locations. Business meetings average 45 minutes, with one person typically online at each location. Conference calls convened for training purposes usually last longer—for an average of 90 minutes—and usually include more people at each location.

Applications

These are averages. In actual practice, audioconferencing can be used for 15-minute updates among three persons or for day-long meetings of thousands. Following are several examples of how audioconferencing is used.

Extra Sales Revenue

R.J. Reynolds cost-justifies audioconferencing in terms of hard-dollar savings and business impact. Hard dollar savings can be measured by money that is not spent on travel, research and development, or it can be measured by an increase in sales. For example, one of R.J. Reynolds' subsidiaries, Kentucky Fried Chicken, was able to get new stores online two weeks ahead of schedule by utilizing audioconferencing. This two weeks of extra sales revenue amounted to about $3 million in savings.

Contract Negotiations and Reviews

Continental Insurance Company's legal department meets with some of its offices and equipment vendors by audioconferencing. The Continental staff people attend these meetings in their own teleconference rooms and the vendors participate at their premises. Before the audioconference, members of the legal department review a contract and send it with their comments through interoffice mail to the appropriate department. Then they conduct an audioconference to review the contract with those departments in various locations in the Northeast. Once all the Continental people are in agreement, they have an audioconference with the vendor to discuss the contract. This type of teleconferencing reduces the length of the contract review cycle, decreases travel and allows everybody to the involved in the contract negotiation.

Continuing Medical Education

Upjohn and the University of South Dakota School of Medicine use audioconferencing to reach physician audiences that Upjohn had not been able to serve previously. In the Midwest, where distance, weather and travel costs can prohibit meeting in person, audioconferencing successfully delivers continuing education programs. Doctors gather at a central location after reviewing printed information provided by Upjohn. Slides are shown during each conference, and a subject expert serves as narrator at the sending location. The participants can ask questions and receive answers from the expert.

These one-hour audioconferences for physicians have proven to be a valuable educational venture for the Upjohn Company. This method of delivering information has been well received by the participants and the planners. It is cost effective and gives Upjohn access and increased rapport with audiences that they previously had not been able to service.

Focus Groups

On the recommendation of a communications task force, The Travelers Insurance Company had to interview a cross-section of field and home office people to find ways to improve communications. This was not difficult to do in the home office—all the company had to do was call a traditional meeting. However, gathering field staff presented a problem. The Travelers conducted a successful pilot audioconference and has since conducted 27 focus groups using audioconferencing to link people in field offices.

Each session lasted about two hours. The participant response was so enthusiastic that the moderators tended to get too involved in the discussion. With additional training, however, the moderators were able to communicate with more than 300 people in just a few weeks, using audioconferencing for the field people and face-to-face meetings in the home office.

Group Discussion

A group of farmers meets via audioconference to discuss the advantages and disadvantages of various brands of fertilizer. Some of these farmers are thinking about buying the product, and others who are using it share their experiences. In this case, the moderator acts as a facilitator for the discussion. In another example, a group of psychiatrists uses an audioconference to discuss a prescription drug that is new and somewhat controversial.

Distance Education

Distance learning is instruction that takes place while the student is physically distant from the instructor. Teleconferencing has helped many educational institutions to accomplish the following:

- A rural high school in Arkansas provides advanced mathematics to a small class whose size does not warrant hiring a teacher.

- High schools in rural Kansas provide college credit courses to honor students.
- Elementary schools in California make sure that all of the students drill in mental arithmetic and spelling at home.
- High schools in a low-income area of Texas obtain the basic skills, as well as advanced mathematics, for a small group of students.
- School officials in rural Iowa make certain that science teachers keep up-to-date in modern physics in terms of both information and teaching techniques.

Tips for Planning an Audioconference

Advance planning is critical to the success of an audioconference. Several guidelines are presented below.

Interconnection Options.

There are several ways of convening an audioconference.

1. Most large telephone systems have some type of three-way or conference call feature. This is usually the least expensive mode of calling, but it has limitations. This mode is best suited to short meetings among a few persons, and is often used for spontaneous meetings.
2. AT&T and other telephone companies provide traditional conference call service on a dial-out basis. The greatest advantage of this service is its reasonable cost. Disadvantages include inconsistency of audio quality; difficulty in rejoining the call if disconnected; and limited to dial-out mode only.
3. A number of companies offer conference call services in competition with the traditional services mentioned above. Special electronic equipment, called an audio bridge, offers a variety of calling features, improved audio quality and operator service. As a rule, audio bridging services charge an hourly service fee per line in addition to long-distance charges incurred by the conference call.

4. Companies with a high volume of conference calls can purchase and install audio bridges on their own premises and in conjunction with their telephone system.

Scheduling Arrangements

Success in a conference call begins long before dialing. All attendees should be notified in writing, if time allows. Such a notice should include the date and time of the meeting expressed in the attendee's time zone; the expected duration of the meeting; the purpose of the meeting; calling instructions with appropriate telephone numbers; RSVP instructions, if any; and a list of expected attendees. A copy of the attendee list, grouped by site, should be retained by the meeting leader for use during the conference call.

Once the date of the meeting has been decided, the length of the call should be set. Most audioconference calls cover material more quickly than face-to-face meetings. However, the most common mistake in planning is to cram too much content into one meeting. A good rule of thumb is to allow 30% to 50% of the allotted time to interaction. Thus, out of every hour of meeting, 20 to 30 minutes should be allotted to discussion and 30 to 40 minutes to prepared content.

Long meetings require scheduled breaks. The first meeting segment should be the longest, as attendees are fresher and are able to concentrate for a longer period of time. The first break should come approximately 90 minutes into the session. After that, breaks should be planned every hour. Typically, breaks last from five to 15 minutes. Lunch breaks must be carefully planned, particularly if many time zones are involved.

Once the date and times for the meeting have been determined, the meeting leader should notify all parties. If a dial-in or "meet me" mode has been selected, the notice should indicate when the parties should dial in, not just when the meeting will start. This time is related to the number of persons on the call. For example, a small call could be announced as follows:

> Annual Conference planning committee will meet Thursday, June 6, 1991, from 2:00 p.m. to 4:00 p.m. Eastern Time (1:00 p.m. to 3:00 p.m. Central Time). **Please dial 201/555-1234 a few minutes before the hour so that we can start on time.**

For a larger meeting of 35 locations and 1500 attendees, the following notice might be sent.

Training Needs Assessment Meeting will be held Thursday, June 6, 1991, in your training auditorium from:

10:00 a.m. to 6:00 p.m. Eastern Time
9:00 a.m. to 5:00 p.m. Central Time
8:00 a.m. to 4:00 p.m. Mountain Time

All sites should call in 30 minutes before the hour to assure a prompt start. The telephone number for this meeting is 201/555-1234.

It is a good idea to make the dialing instructions stand out in your memo, either by underlining or through the use of bold face type.

Once the length of the program has been determined, the agenda or program outline should be developed.

Next, make a list of those who are best qualified to present the selected topics. Try to include more presenters than you might choose for a face-to-face meeting. The faceless environment of audioconferencing requires more variety to attract and hold the attention of attendees.

Plan the agenda in short segments. Break down complex topics into five- or 10-minute segments. Intersperse these segments with discussion or questions. *Allow no presenter to speak for more than 10 minutes without a significant change element.* In general, persons can listen to a disembodied voice for no more than 15 minutes without losing attention. Give each presenter 10 minutes to speak. Even if a speaker runs over, he/she will not likely exceed the 15-minute maximum.

Conclusion: Preparing for the Audioconference

Well before the meeting begins, the chairperson and each site coordinator should complete the following checklist:

- If the call requires a conference room, reserve it.
- If special speakerphone equipment is needed, arrange for it.
- If you are unfamiliar with the equipment, practice using it.
- Arrange for coffee or other refreshments.
- Plan no appointments before the meeting that could run late.
- Notify your company switchboard, your secretary and co-workers that you are in a telephone meeting and are not to be interrupted.

MEDIA FOR BUSINESS

- Put a sign on the meeting room door that says, "Teleconference in progress. Please enter quietly."
- Enter the room well in advance and check for dial tone on the equipment.
- See that handout materials are duplicated and in place.
- Check for needed audio/visual equipment.
- Check seating so all attendees have easy access to the microphone.
- Have a copy of the attendance list, the agenda, all pertinent telephone numbers and blank paper for your own use.

AUDIOGRAPHIC CONFERENCING

Introduction

An "audiographic conference" combines an audioconference with visual images. Transmitting images electronically during an audioconference is nothing new; electronic "blackboards" and tablets of various sorts have been used for years. (Some would even consider the addition of fax machine transmission to the audioconference as meeting the definition of an audiographic conference.) The old technologies pale, however, compared to today's PC-based systems with their capabilities for displaying a wide range of visual images and real-time annotation, making possible lively presentations, interactive training and captivating educational programs.

Audiographic Conferencing Systems

The term "audiographic" applies to a family of technologies that use ordinary telephone lines for two-way voice communication and transmission of graphic materials. Every audiographic system incorporates an audioconferencing component that can range from a simple speakerphone to a sophisticated system employing loudspeakers and voice-activated multi-directional microphones.

One of the earliest audiographic conferencing systems in popular use combines the audioconferencing unit with an electronic backboard for the transmission of graphic materials. Using a special pen, an individual at one site writes and draws on the electronic blackboard. The graphic elements are then transmitted over a telephone line. They appear almost simultaneously at the receiving location. Individuals at that location can also use the blackboard at their site to help them explain a particular problem or issue. What they write is displayed at the other site. Two telephone lines are required: one for voice transmission and the second for the transmission of graphics. All the graphic materials are displayed in black and white.

Still video images can also be sent to other locations via telephone lines using a transceiver. At the receiving end, a second transceiver displays the video image on a monitor. The transmission time varies as a function of the resolution, color and complexity of the image—the time may be as short as 10 seconds.

An image storage unit can be used to facilitate a slide show presentation format. Each component of this system—the visual and the audio—requires its own telephone line. Still-frame video systems (also known as "slow-scan" or "freeze-frame") are sometimes used in conjunction with the electronic blackboard. The addition of a special switching device allows the user to change from one to the other to avoid the need for a third telephone line.

Advances in computer technology have led to the development of fully integrated PC-based audiographic conferencing systems that are becoming increasingly popular. These systems combine voice, data, graphics and, in some cases, color still-frame video to create a powerful communications tool. The PC is equipped with conferencing software, special boards to support a graphics tablet and pen, and the video component. One such system uses a half-duplex modem that can simultaneously transmit graphic images, computer data, still video images or voice on a single dial-up telephone line. Others require separate telephone lines for computer data (including graphics) and voice.

Each site on the network is equipped with a PC, graphics tablet and pen, modem, computer and/or video monitor, and audioconferencing unit. Users can prepare CGA-level computer graphics and full-color still video images in advance and store them on disks for transmission in real time. Or, they can transmit them before a meeting or class over high-speed modems, or make copies on floppy disks and send them to each site via mail or courier. Multiple screens of images can be stored in a single directory and called up, either in sequence or randomly, within seconds by any site on the network. Individuals at any site can use the graphics tablet and pen to point to portions of the images that are on the screen or to write on them for simultaneous viewing at all sites.

Applications

Audiographic conferencing technologies are being used in a variety of settings, including educational institutions, corporations, hospitals and the military.

Educational Institutions

More than a dozen high schools, colleges and universities in North America currently use audiographic networks. These systems deliver classes in mathematics, history, computer science, English, economics, engineering, nursing and other subjects to students at remote sites. In most cases, these institutions were faced with the dilemma of how to deliver instruction to small numbers of geographically dispersed students on very limited budgets. To meet the demands of students and the communities they serve, these schools adopted computer-based audiographics systems to extend their educational resources.

Corporations

There are several companies that use audiographic conferencing technologies to augment audioconferencing. AT&T, for example, has one network of more than 20 sites to deliver training to its employees over the system. Employees participate in the training sessions without leaving their place of work, which saves both time and travel expenses and makes effective use of the trainers.

In general, corporate trainers use the system in much the same way as educational institutions. Corporations, however, use audiographic systems for other purposes as well. Pratt and Whitney Canada uses this technology for project management. Project team members from different locations throughout Canada meet daily via audiographic conferencing, something that would not be possible if everyone had to travel to a central location.

Science Applications International Corp., headquartered in La Jolla, CA, has audiographic conferencing links with four of its U.S. sites. They use the system to facilitate decision making among their operating divisions and to link some members of their Science Applications Management Council for daylong meetings four times a year. With the audiographic link, no one needs to miss the sessions when travel time is not available.

The Ford Motor Co. makes extensive use of audiographic conferencing. One of the more interesting applications involves using the still-frame, or slow-scan, feature to send video images from construction projects. Engineers can monitor the projects without traveling to the sites.

In addition to its customer training network, AT&T has three additional audiographic networks that range in size from 50 to 160 sites. One network is used exclusively for sales and marketing, another for operations and the third is a general business network that is used to introduce new products.

Unisys Corp. uses audiographic links with 14 of its subsidiaries on the Pacific Rim. The company conducts business meetings and introduces new products via the network.

Hospitals

Specialists in a Boston hospital use a computer-based audiographic conferencing system that can store and display high-resolution video images to provide pediatric radiology consultation to physicians in rural settings. Video images of the X-rays are transmitted to the consulting physician, who uses the tablet and pen to point out particular features and abnormalities.

Many medical specialists are exploring the use of audiographic technology to train and consult with healthcare providers in developing countries. In the spring of 1988, for example, a computer-based audiographic system delivered a series of lectures on cell biology from the Boston University School of Medicine to a classroom in mainland China.

The Military

In sheer numbers of students, training for the Armed Forces surpasses both educational institutions and corporations. The military must train and retrain hundreds of thousands of learners on a regular basis. The U.S. Army Materiel Command is beginning to use audiographic systems to assist them in this task. A typical class consists of about four hours of live televised instruction, followed by another two hours of interaction using the PC-based audiographic system.

Benefits and Limitations of PC-Based Audiographic Conferencing

There are some limitations to using audiographic systems for delivering training and education. The visual information is less active than with a full-motion video system. Participants who are not highly motivated to learn the information that is being presented will not have their attention captured by moving and rapidly changing images. Audiographic systems are very effective, however, for highly motivated users who are willing to take an hour or so to learn to use the technology.

Successful users of audiographic delivery—both instructors and students—need to be trained to use the system and should practice with the equipment. Most instructors feel comfortable and perform competently with a few hours of practice as long as the technical considerations of the systems are explained clearly to them. It takes about an hour to train a group of up to 20 students who will be using an audiographic system. Successful use of this technology also requires advanced preparation of the materials to be used during the session.

For all these limitations, there are enormous benefits to using the audiographic alternative when face-to-face communication is not possible. In an instructional setting, a single teacher can interact simultaneously with either large or small groups of remote students at many different sites.

There are also some less obvious advantages to these PC-based audiographic conferencing systems for instructional applications.

Most people who are accustomed to working with computers pay very close attention to the computer screen and are used to getting complex information from that source. As a result, students who are asked to watch a computer screen to get their course-related information find it very easy to do so. Also people working with computers are used to reacting and being active. For instance, they push keys or move a mouse in response to instructions they get from the machine, and the machine responds in turn. This type of interaction is desirable in a learning environment.

Faculty and students who have used both audiographic conferencing and instructional television find that interactions and personal information exchanges are easier with audiographic conferencing. Both students and faculty who use the system well testify that the audiographic conferencing system allows an intimate relationship to develop.

Because audiographic technology is not well known to faculty and students, they are willing to be trained to use it—and the special techniques for teaching and learning at a distance. Faculty interviewed for a study by the University of Maryland University College indicated that their teaching improved as a result of presenting a course using audiographic conferencing. They were better prepared and began to think about creative ways to use the class sessions.

Finally, an audiographic conferencing system can be inexpensive and flexible. The cost of such a delivery system is relatively modest. Each unit is portable and can be moved from place to place easily and set up in less than an hour. The only communications connection required is a telephone jack, and telecommunications charges are the same as for a standard telephone call (unless there are multiple sites on the system and a telephone bridge is used, resulting in additional telecommunications charges). In addition, the equipment itself does not have to be completely dedicated to audiographic conferencing. When not being used for classes, the PC can function as a regular work station. This versatility is another way to save money on a conferencing system that allows for real-time exchanges of voice, data, graphics, and still-frame video.

VIDEOCONFERENCING

Introduction

The videoconference easily wins as the most intriguing of the teleconferencing options. It probably also wins as the least understood and most maligned option, by those who believe, erroneously, that it is a substitute for, rather than a supplement to, the face-to-face meeting.

Over the past decade, the videoconference has endured wild market forecasts and predictions of doom to maintain a steady growth rate marked by some truly innovative technological developments.

Dick Tracy used videoconferencing to bust up crime. The staff on the moon, in the movie *2001*, used it to keep in touch with families and hold conferences with colleagues on earth. Science fiction? Not entirely. While we don't yet have earth-to-moon or wristwatch videoconferencing, there are plenty of down-to-earth uses of digital videoconferencing which are proving to be essential to business, government and educational institutions.

Corporate Videoconferencing Applications

Corporations are the heaviest users of videoconferencing. Whether they have elaborate systems installed in boardrooms or use desktop terminals, most corporations hold videoconferences, either routinely or for special purposes. And many organizations find videoconferencing to be a strategic competitive tool.

Examples of regularly scheduled meetings include:

- Project management
- Intergroup coordination

- Management meetings
- Information dissemination

In addition, special-purpose meetings can be held to handle a crisis, explore unusual opportunities or present information on special situations.

Most organizations still justify the initial and ongoing expense of installing and operating videoconferencing networks on the basis travel dollars saved. But the real value of videoconferencing is that it saves time and traveler frustration and allows more timely decision making. These observations are supported by the experience of videoconferencing organizations.

Banking

AMSOUTH Bancorporation is growing internally and through acquisitions. To speed up commercial loan approvals and reduce redundant travel, a multipoint videoconferencing network was installed at major bank facilities. The system permits automatic mixing of the audio for all sites and switching of video images during the videoconference. The system allows loan officers to be more responsive to the bank's clients. Via videoconference, the loan officers receive input from appropriate bank executives without traveling to a central site to meet with the bank's Credit Review Committee.

In addition to the credit review function, other committees, such as the asset and liability committee and the Board of Directors, meet on a regular basis using video conferencing.

The bank's acquisition program and its introduction of new financial products has mandated extensive training programs for loan officers and other employees. Holding classes simultaneously at major locations allows for faster, more effective training programs.

The videoconferencing network also allows timely dissemination of critical information to bank employees and large investors. On Black Monday, October 19, 1988, for example, the president of the bank used the network to communicate its financial position and reassure depositors.

Retailing

Sears is expanding through vertical and horizontal acquisitions. The chain installed a major multipoint network of videoconferencing rooms to improve communications among its regional offices. The network was later expanded to include major store locations. Although Sears initially used videoconferencing to enhance brand and product management discussions, the network now supports multiple functions including distribution, accounting, budgeting and employee training.

Engineering

For several years, the Ford Motor Co. has used videoconferencing on a daily basis to link design engineers, located in several countries, with testing facilities and production plants.

Design and production engineers view and discuss equipment, parts, drawings and blueprints with key people at each location. Graphics displays and hard copies of drawings allow modifications to be made during the videoconferences. Design and production personnel can work together without unnecessary travel, and, even more important, engineers can access critical information resources via videoconferencing.

Telecommunications

One example of inter-carrier videoconferencing is a series of meetings held between AT&T and British Telecom. Over a six-month period, the two companies discussed engineering, product management and product promotion relative to the conversion of a submarine cable to a satellite route. Besides avoiding several trips involving eight to ten people, the series of videoconferences allowed a smooth, on-time transition to the new route with no interruption of service.

Insurance

Aetna realized significant cost savings with its first videoconferencing installation. Video conferences between two locations less than 10 miles apart cost-justified their system within the first year of operation—through time saved by employees avoiding the short commute between the two sites.

Medical Videoconferencing Applications

Hospitals and clinics use videoconferencing to evaluate patients, exchange information with subject matter experts and train personnel.

The Mayo Clinic uses videoconferencing for educational and clinical healthcare applications. Physicians at the clinic in Minnesota are linked with physicians in the Mayo Clinics in Florida and Arizona. The conference room in Rochester, MN, features five cameras and videotaping and slide transmission capabilities. Examination rooms are equipped with cameras and monitors. The system is used for consultations and supports a variety of video diagnostics like X-ray, telepathology and ultrasound tests.

The Bethesda Lutheran Home, which specializes in mental retardation services, uses videoconferencing to link its headquarters in Watertown, WI, with facilities in Shawnee Mission, KS. The Home uses videoconferencing for courses, Board of Directors' meetings, job interviews, staff consultations and demonstrations for college students.

Government Videoconferencing Applications

The U.S. government is the largest single user of videoconferencing in the world, and many local and state governments are rapidly expanding their use of the medium. The uses of videoconferencing within the government vary as widely as each organization's tasks. A few examples follow.

U.S. Army Materiel Command

The AMC requisitions, tests and distributes materials. A videoconferencing network has been established to allow decision makers to discuss major contracts and programs.

Four-star generals conduct regular secure meetings with suppliers to make purchase decisions. Materiel distribution, personnel deployment and project management are other subjects discussed on the videoconferencing network. The decision-making process is now quicker and more efficient and, according to reports, each videoconference creates an operational cost savings of about $1000.

U.S. Naval Underwater Systems Command

Developing efficient submarine combat systems requires extensive meetings and close coordination among various technical and administrative personnel. The research and development group at NCSC uses its videoconferencing network to accomplish its mission with less chance for miscommunication and with better coordination.

National Aeronautics and Space Administration

NASA manages the launches of the most complex pieces of equipment built in the United States today. One of NASA's objectives is to link private sector contractors with NASA engineers and flight control administrators.

With videoconferencing, NASA personnel can manage complex projects from videoconference rooms across the country. Videoconferencing facilitates the multiple, detailed checks and verifications necessary to NASA's activities.

Local Justice Departments

Local governments also use videoconferencing to expedite procedures. The Brooklyn District Attorney's office, for example, has tested videoconferencing as a faster way to gather statements from victims and witnesses and to speed up the arraignment process of arrested individuals. More important, videoconferencing has increased (from 10% to 90%) the number of victims and witnesses who agree to testify. As a result, the Manhattan District Attorney's office, among others, is exploring the use of videoconferencing for similar applications.

Education Videoconferencing Applications

Pennsylvania State University

Credit courses, student interviewing and staff development training are three ways that videoconferencing is used by Pennsylvania State University.Two of the three sites operated by the university are separated by more than 200 miles (University Park and Erie). Students at remote sites can receive a wider variety of courses, since they can access professors at both locations. The third site, located at Hershey Medical Center, gives students easier access to medical courses.

University of Missouri

University of Missouri professors spend more time in their classrooms and less time on the road thanks to a videoconferencing system that allows two-way communication between remote classrooms. The university installed a new facilities network and digital videoconferencing systems to carry voice, data and video among four campuses.

Students at remote sites have matched the performance levels of their on-site counterparts. The university finds videoconferencing to be a valuable teaching tool. Students can take courses that were previously unavailable because professors can (literally) be in two places at the same time.

Videoconferencing Technology

The slow but successful growth of videoconferencing is due to technical advancements not available in the 1960s or even the 1970s. Picturephone,™ premiered by Bell Laboratories at the 1964 World's Fair, didn't sell. Videoconferencing through the telephone network could not be made economical with analog technology. To accommodate human speech, telephone companies elected long ago to restrict the bandwidth of a voice channel to a maximum range of 4000 Hertz (Hz) (cycles per second).

An analog video channel requires a tremendous amount of bandwidth to transmit pictures. The range of frequencies needed to reproduce a high-quality motion TV signal is at least 4.2 million Hz (4.2 mHz). Bell engineers were able to reduce Picturephone's bandwidth requirements to 1 mHz, but this was not enough to make transmission economical. In the 1970s, conversion of the analog video signal to a digital bit stream enabled the first significant reductions in video signal bandwidth.

Digitizing a video signal produces data at a rate of 90 million bits per second (90 mbps). This large amount of data is generated because each frame of a television picture consists of more than 300,000 picture elements (pixels), each pixel requiring eight bits to adequately describe its color and intensity, and because a television picture refreshes itself at the rate of 30 frames per second.

Video Compression

A typical television picture (particularly of a business meeting) does not change too rapidly, and adjacent pixels are often nearly identical. In the 1970s, engineers realized that data compression techniques could be applied to reduce picture bandwidth. Video compression techniques greatly reduce the amount of data needed to describe a video picture and enable the video signal to be transmitted at a lower, and less expensive, data rate.

The device used to digitize and compress an analog video signal is called a video codec, short for COder/DECoder. A codec converts an analog video or audio signal into a bit stream for transmission, then reconverts the signal to analog at the remote site. Codecs manufactured by one vendor are generally not compatible with those made by another vendor.

As the codec digitizes the signal, it also compresses it. The biggest commercial breakthrough came in 1982, when several codec vendors demonstrated that acceptable compressed video could be transmitted at 1.5 mbps, or a compression ratio of 60:1 A few years later, compression to 56 kbps (a single digitized telephone line) had been achieved by several codec manufacturers.

Reduction of transmission rate requires trade-offs in picture quality. As the transmission rate is reduced, less data can be sent to describe picture changes. As a general rule, lower data rates yield less resolution and less ability to handle motion.

Some persons use the phrase "full motion video" to describe video that has been compressed by a codec, in contrast with the limited or still frame video sent by audiographic equipment.

Digital video compression allowed videoconferencing to become, if not inexpensive, less cost-prohibitive. Videoconferencing has achieved a synergy of lowered transmission costs, improved price-performance of video compression technology and integrated products. While videoconferencing is not cheap or ubiquitous, it is maturing.

Videoconferencing Networks

Organizations wishing to use motion videoconferencing have several options: to build in-house facilities, to use public videoconferencing rooms or to lease time on other private company networks (the last option is less common).

Several public networks are currently in operation. Examples include US Sprint's The Meeting Channel and the AT&T Accunet network.

Public rooms have many advantages. For an organization considering in-house videoconferencing facilities, public rooms offer a good test ground without capital investment risk. Infrequent users may find it more economical to pay for a public room than to bear the costs of a private network.

Public rooms are located across the United States and worldwide. Trained staff can answer questions or help prepare business persons to use the rooms. Gateways are available to translate video signals between incompatible codecs.

However, public rooms have drawbacks. Business people must travel to the videoconferencing room. The meeting takes place in an unfamiliar environment. And companies that are regionally headquartered in small or rural areas may not have easy access to public rooms

A company has no absolute control over the availability of a public room. If a corporate emergency occurs and the public room in a city is already booked, the company can't "bump" the reserved user.

Guaranteed availability is often the primary reason that corporations build private networks. Companies also like to keep control over usage, access and security. The network can be designed to reflect the corporate culture and image. And the budget for videoconferencing can be planned in advance.

Videoconferencing Systems

As prospective users quickly discover, videoconferencing includes the products and services of several different industries.

A videoconferencing project usually begins with an internal study and cost estimate. This analysis may be done by an outside consultant or by an employee of the user organization.

The equipment (cameras, microphones, etc.) is usually obtained from a "systems integrator," a company that designs and engineers the videoconferencing system and provides and installs the audio and video equipment. Several companies provide systems integration geared directly to videoconferencing.

Transmission links (analog or digital) are normally leased from a common carrier. Some organizations will lease full-period digital channels to carry compressed video signals; others prefer to lease channel time on a pre-reserved, hourly basis. If an organization has its own telecommunications network, it may choose to multiplex videoconferencing signals onto the network.

The video codec can be purchased directly from the manufacturer, through the systems integrator or through the common carrier.

Connecting the Systems

During a videoconference, audio, video and data signals are transmitted between sites on a single combined channel or on separate channels. (Audio is sometimes transmitted over a dial-up telephone line.) The transmission channel can be analog or digital; signals can be sent by satellite, microwave, terrestrial circuits or a combination of these.

Most business television point-to-multipoint meetings are transmitted on full-bandwidth analog video channels (identical to broadcast television channels). Most videoconferences, however, run with compressed video signals on digital transmission links.

With satellites, geographically distant sites can be linked as easily as nearby ones. Land-based transmission is usually priced according to distance, but satellite transmission is insensitive to the mileage between locations. Point-to-multipoint videoconferencing is also easy, since an unlimited number of locations can receive a satellite video signal. A third advantage of satellite transmission is that the leased channel can be swapped among many locations.

Satellite transmission also has drawbacks. An electronic signal uplinked to the satellite and downlinked to the receiving site makes a round trip of 44,600 miles. Even though these signals travel at the speed of light, the round trip is long enough to cause a delay of almost half a second. In a videoconference this delay, coupled with the codec processing delay, creates a momentary pause between the time a person makes a statement and the persons at the remote end react to it.

Fiber optic transmission is a growing alternative. A single fiber optic cable can transmit over a billion bits per second, enough for plenty of simultaneous videoconferences and other data traffic. Across the United States, major transmission carriers have already connected high-traffic routes with fiber optic cabling and are bringing fiber to businesses and residences. Many institutions are building private fiber optic transmission networks to carry voice, data and video.

A fiber optic cable can carry analog or digital video signals. The huge bandwidth available on a single fiber allows broadcast or near-broadcast quality video signals to be transmitted economically. Digital cross-connects, similar to telephone company voice switches, are used to create end-to-end transmission paths over fiber optic cables. They can also emulate the point-to-multipoint feature of satellite transmission by multiplying a videoconferencing signal and transmitting it to many sites.

Videoconferencing signals can also be sent by digital or analog microwave systems, or on dial-up digital transmission lines. AT&T and the Regional Bell Operating Companies offer dedicated and switched 56 kbps digital telephone lines.

Regardless of transmission method or speed, an organization must decide whether to lease a full-period channel or to lease transmission on demand.

Leasing a full-time channel is more economical for frequent usage, and it guarantees instant, constant access to the videoconferencing network. On-demand usage is better when an organization is experimenting with videoconferencing or uses it infrequently. And of course, a combination may be desirable: full-time leased channels to conference among domestic locations, and occasional-use channels for less frequent international meetings.

Multipoint Videoconferencing

The majority of videoconferences connect two locations (point-to-point) using a single transmission channel between public and/or private rooms.

When organizations wish to connect more than two locations in a single conference, additional considerations must be addressed. Except for conferences involving three sites, economic and technical limitations require a few compromises in multipoint electronic meetings.

The simplest method, a point-to-multipoint videoconference in which a single site broadcasts to multiple sites, is addressed in the next section of this chapter.

BUSINESS TELEVISION

Introduction

The teleconference technology with the newest name, business television, saw a 100% increase in market size in 1988. By all indications, it will continue to grow as organizations find it an effective medium for creating and distributing training, information and motivational programs to employees and, increasingly, business television is being used to reach out to customer and other external audiences with innovative market positioning programs. To clarify the formats and services that comprise business television, a group of experts developed the matrix of definitions shown in Table 13.2.

U.S. companies have been using television as a training and information tool for almost 40 years. In 1987, more than 7500 companies in the United States had their own in-house production facilities, a 114% increase in five years. While the majority of television programs produced for and by the business community are duplicated and distributed on videocassettes, increasing attention has been paid in recent years to newer, more efficient distribution systems. In the early 1980s, satellite delivery made corporate TV networks possible and fueled the growth of a new kind of teleconferencing known as business television (BTV).

TABLE 13.2: BUSINESS TELEVISION NETWORKS AND SERVICES

BUSINESS TELEVISION is the production and electronic transmission of broadcast video programming to targeted user groups in corporations, educational institutions, government agencies, and other non-profit organizations.

	PRIVATE NETWORK	PROGRAMMING NETWORK	EDUCATIONAL NETWORK	PROGRAMMING SERVICE
Audience	Employees or franchisees of the network owner, or customers or resellers whose primary relationship with the network owner is the purchase of goods or services *other than* programming.	Wide range of viewers, who are employed by or are otherwise affiliated with an organization whose primary relationship with the network owner is the purchase of programming.	Students, professionals, or community members who register or enroll in a course of study that consists of one or more broadcasts.	Individuals who purchase programming from the network owner; or a wide range of viewers who are employed by or otherwise affiliated with an organization whose primary relationship with the network owner is the purchase of programming.
Nature of Programming	Programming specific to the network owner, including information, training, and motivation to improve job knowledge and skills, to build employee morale, and to encourage customer or reseller loyalty.	Programming tailored to a vertical industry segment (e.g., automotive, banking, insurance) or a subject matter interest area (e.g., management, computer technology) or a market segment (e.g., shoppers).	Instructional material delivered to meet specific course objectives, with the majority of the programming part of an established curriculum leading to primary or secondary school grade level credit, an undergraduate or graduate college degree, continuing education credit, or professional certification.	Programming created for subject matter interest areas (e.g., contemporary business issues and practices, management skills), designed for broad appeal across a variety of market segments.

TABLE 13.2: BUSINESS TELEVISION NETWORKS AND SERVICES (Cont.)

	PRIVATE NETWORK	PROGRAMMING NETWORK	EDUCATIONAL NETWORK	PROGRAMMING SERVICE
Program Schedule	Both regularly scheduled and occasional programming. Frequency varies by network.	Frequent, regular programming, primarily on a daily or weekly basis.	Regularly scheduled programming. Frequency varies.	Both regularly schedule and occasional programming. Frequency varies by programmer.
Viewing Fees	Generally none, since capital costs and expenses are borne by the network owner; in some cases, particularly with franchise organizations, customers, and resellers, a portion of costs may be assessed to viewer sites.	Generally offered on a subscription basis with fees for equipment and programming assessed per company, site, or viewer for on-going programming services; some one-time pay-per-view arrangements; some advertising-supported networks with no viewing fees.	Covered by course enrollment feels; or provided as free community services; or operated by a school district or state-wide or regional service with funding through tax dollars.	Subscription basis, charged per company, site, or viewer for one or a series of programs; some advertising-supported programming.
Viewing Sites	Reception equipment owned or leased by the network program provider or its viewers and generally located on viewer premises.	Reception equipment owned or leased by the network program provider or its subscribers and generally located on subscriber premises.	Reception equipment owned or leased by the network program provider or viewer; or other facilities rented by the program provider on an as-needed basis.	Public facilities rented by the program provider on an as-needed basis; or permanent or transportable equipment owned or leased by the viewer.

Source: KJH Communications. Developed by Jeff Charles, Kathleen Hansell, Susan Irwin, Richard Neustadt, Virginia Ostendorf and Doug Widener.

Business television uses communications satellites to broadcast informational television programs from a single location—often a company's corporate headquarters—to a targeted audience, located at multiple, geographically dispersed locations. The programs are often broadcast live, giving the audience the opportunity to phone in questions.

Between 1982 and 1988, more than 50 of the largest and most successful corporations in the United States, representing a wide range of industries, installed satellite television networks to enhance their internal communications. Hundreds of other organizations have taken advantage of this medium to broadcast special events to targeted audiences.

Business television has emerged as a powerful tool for companies that need to deliver timely information to a large, dispersed audience.

Benefits of Business Television

Some benefits of business television are quantifiable, such as savings in travel dollars or duplication of tapes. Most are "soft" benefits, such as the ability to deliver time-critical messages immediately and get instant feedback from the field. The benefits of business television can best be described by example.

Immediacy

The ability to transmit live television by satellite allows time-critical information to reach a targeted audience immediately.

During the week following "Black Monday," October 19, 1987, Merrill Lynch used its private network every day to apprise its stockbrokers of up-to-the-minute information about the market conditions. In addition to the obvious benefit of immediacy, the brokers were given a sense of security by remaining in contact with corporate headquarters during a time of fear and uncertainty. Merrill Lynch network manager Marilyn Reed said that the cost of the 475-site video network was justified by its use during that single week.

Simultaneous Message Delivery

When Digital Equipment Corp. established a significant change in pricing and pricing policies, it needed to get the information to the sales force, quickly and with a consistent message, before it went to the customers. An ad hoc network of receive sites was established at 80 hotels around the country, and two days of panel discussions, interspersed with question-and-answer sessions, were broadcast to more than 8000 employees.

Feedback from the Field

There is no substitute for the immediate feedback made possible by live television with questions phoned in from the receive sites to the studio.

The U.S. Army installed a network to augment its live classroom instruction for required courses on logistics management. Using an audio bridge and push-to-talk microphones, the lines are kept open throughout the daily two-to-three-hour classroom sessions. This system allows for continuous interaction. The students can ask questions as well as participate in open discussions with the instructors and with students at other receive sites.

Greater Access to Experts

Retired Admiral Grace Hopper, the first female Navy admiral and the developer of the Cobol computer programming language, is a consultant to Digital Equipment Corp. Admiral Hopper lives in Washington, DC, and her age and health limit her travel. Digital's business television network has brought her to more Digital employees than could ever have been possible before.

Efficiency of Training

In addition to running full-time residential training centers, companies like Aetna Life and Casualty send their training programs on the road. But there is no way to reach all of Aetna's claims adjustors with time-critical information about quickly changing insurance regulations and the products the company is representing. Aetna's satellite network is alleviating that problem by broadcasting training programs to its geographically dispersed personnel.

Travel Reduction

Savings in travel dollars is still used as one of the primary justifications for installing a permanent network. In addition, staff time saved by not traveling is measurably significant. The Army has found that the cost of delivering a three-week course over its satellite network is $500 per student, as opposed to $2000 to deliver the same course at a residential facility. Over $7 million has been saved for the U.S. taxpayer since the Army's Satellite Education Network was installed in 1984.

Motivation

The A. L. Williams Co. sells more life insurance than any other insurance company in the United States. Every Monday morning, Art Williams, the chairman of the company, motivates his sales force of 140,000 agents via live broadcasts on the 1000+ location ALW-TV network. The agents also receive about 20 hours per week of training and informational programming transmitted from the studio at the A. L. Williams headquarters in Atlanta.

Increased Productivity

The most significant benefit of business television can be summarized by the expression, "the whole is greater than the sum of its parts."

Immediacy, simultaneity, feedback from the field, access to experts, more efficient training, reduced costs and increased motivation all lead to increased productivity. Used appropriately, business television can improve communications and, regardless of the industry or the application, better communications leads to greater productivity.

Business Television Applications

Business television users have identified a variety of applications that fit their needs. Most of the programs are developed and produced by the companies themselves. They use in-house studio facilities, if available, or contract with commercial production companies. Some of the key applications are described below.

Training

The most widely used application for business television is training—sales training, management and technical training, training for support personnel and, increasingly, customer training.

Digital Equipment Corp. has developed a monthly program series for the Digital Video Network (DVN) on competitive selling, with a very focused topic and audience. Rebecca Warshawsky, DVN manager, says that the series has been "extremely effective in shortening the sales cycle" and that it has actually resulted in closing sales. Ms. Warshawsky calls the series the "flagship program" for the network. "It was the right application at the right time for our position in the industry," she said. "Programs on strategy and competitive positioning strike a chord with our sales staff."

Domino's Pizza has installed a satellite network at its major distribution centers and is using it for employee training. Subjects range from making pizza dough to improving customer service. Domino's is also making its network available for adult literacy training and the delivery of college courses.

Employee News/Information

Keeping people informed about their company and their industry becomes more difficult as the company grows and becomes more geographically dispersed.

Federal Express broadcasts a morning news show over FXTV, a business TV network that now reaches 800 Federal Express locations and is being installed at additional sites. "FedEx Overnight" provides employees with a five-minute daily report, including weather problems and flight schedules, as well as feature stories on various employees, offices and worldwide events. Ford Motor Co. uses the 280-site Ford Communications Network to deliver a daily company news program, as well as special events, training programs and broadcasts serving a variety of other functions.

Product Announcements

Introducing new products effectively—to a sales force, service personnel or customers—is time-consuming, expensive and critically important.

By renting hotel facilities where satellite antennas have been installed, or by bringing transportable "dishes" to hotels or directly to corporate sites, organizations reach a wider audience and obtain a more powerful impact. General Motors, for example, introduced its 1987 car models at a satellite teleconference, augmenting coverage in the company newsletter, trade publications and the national press.

For many companies that have installed private networks, new product announcements and product updates constitute regularly scheduled programming. Merrill Lynch uses its network weekly to announce and discuss new products with its brokers. J.C. Penney, with hundreds of store sites installed, displays its new clothing in broadcasts for buyers.

Press Conferences

When the Ford Motor Co. was on the verge of settling its 1987 collective bargaining disputes with the United Auto Workers, the Ford Communications Network promised to provide live coverage of the results to its employees over the company's satellite network. Finally, a deal was struck and the handshake between the president of the UAW and the chairman of Ford was seen live by the people most affected by it. Said Jack Caldwell, director of internal communications for Ford, "The number one question for a Ford employee was 'Are we going to have a contract or are we going to strike?' Most workers' lives would be governed by this result . . . For once, the employees of the company heard the results of the bargaining agreement at the same time as the press."

Special Events

Since the early 1980s, satellite downlinks have been installed at hotels, hospitals and universities that make their facilities available for videoconferences targeted to specialized audiences.

Texas Instruments Inc. used videoconferencing as early as 1980 to hold its annual stockholders' meeting. Since then, in addition to using business television for employee and customer meetings, Texas Instruments has held two major educational symposia on artificial intelligence. The first symposium was broadcast to 500 locations and the second to 850 locations in 16 countries. Total attendance at the two events, which were free, was 80,000 people. Texas Instruments now has installed a permanent network.

Cross-Networking

Two of the attractive features of scrambled, addressable satellite networks are the abilities to control which sites in a network receive any given program and, of course, to issure that no one from outside the company can access that signal. These capabilities also make it possible to share programming with other networks when appropriate.

Cross-networking has been used by Hewlett-Packard, Texas Instruments, ComputerLand, Digital Equipment Corp., Eastman Kodak and others for presentations to customers, sharing of experts' advice and a variety of other purposes.

External Programming

The proliferation of private networks and individual TVROs (television receive-only antennas) has spawned a growing number of satellite-delivered program services. These are aimed at specific industries or at targeted training and educational needs that cross company boundaries. Program sources fall into two categories: programming networks and satellite-delivered program services. Programming networks include a subscription to regularly scheduled programs, as well as downlink hardware and installation. Individual programs or series of teleconferences are available (usually for a fee) to any location that has an installed satellite downlink or that rents a portable downlink to receive the program. These programs services target private corporate networks and use hotels, hospitals and universities to reach their intended audiences.

Conclusion

Though the number of organizations that have installed their own private satellite networks is still relatively small—56 as of February 1989—that number has grown from only three in 1982 and is projected to grow to more than 300 by 1992. The current number of permanently installed locations to which they broadcast is close to 12,000. Between 1983 and 1987, the number of broadcast hours of business television increased from 500 to 10,000.

Business television, though still in its infancy, has demonstrated its ability to meet many significant corporate communications needs.

FURTHER INFORMATION

Periodicals

Communications Consultant
Jobson Publishing Corp.
352 Park Ave. S.
New York, NY 10010-1709

Communications News
7500 Old Oak Blvd.
Cleveland, OH 44130

Communications Week
CMP Publications
600 Community Dr.
Manhasset, NY 11030

Communique
Int'l Communications Assn.
12750 Merit, Ste. 710, LB-89
Dallas, TX 75251

Corporate Video Decisions
295 Madison Ave.
New York, NY 10017

**CPE Strategies and
   Telecommunications Product Review**
Marketing Programs & Services
  Group, Inc.
P.O. Box 217
Gaithersburg, MD 20877

Data Communications
McGraw-Hill, Inc.
1221 Ave. of the Americas
New York, NY 10020

Network World
CW Communications, Inc.
375 Cochituate Rd., Rte. 30
Framingham, MA 01701

Satellite Communications
6300 S. Syracuse Way
Englewood, CO 80111

Satellite News
Phillips Publishing, Inc.
7811 Montrose Rd.
Potomac, MD 20854

Tel-Coms
2801 International Lane
Ste. 205
Madison, WI 53704

Telecom Highlights
P.O. Box 1609
Paramus, NJ 07653

Telecommunications
685 Canton St.
Norwood, MA 02062

Telecommunications Week
Business Research Publications
1036 National Press Bldg.
Washington, DC 20045

**Telecommunications Products &
   Technology**
PennWell Publishing Co.
One Technology Park Dr.
Westford, MA 01886

**Telephone Engineer & Management
   (TE&M)**
124 S. First St.
Geneva, IL 60134

214   MEDIA FOR BUSINESS

Telespan Newsletter and
Telespan's Business TV
c/o Telespan Publishing Corp.
P.O. Box 6250
Altadena, CA 91001

Via Satellite
P.O. Box 2000-141
Mission Viejo, CA 92690

BOOKS

Electronic Meetings, Robert Johansen, Jacques Vallee, Kathleen Spangler Vian (Addison Wesley, 1979).

Teleconferencing and Beyond, Robert Johansen (Data Communications/McGraw Hill Publications Co., 1984).

Teleconferencing: Linking People Together Electronically, Kathleen Kelleher and Thomas B. Cross (Prentice-Hall, Inc., 1985).

Teleconferencing: Maximizing Human Potential, Robert Cowan (Reston Publishing Company, 1984).

Teleconferencing Technology and Applications, Christine Olgren and Lorne Parker (Artech House, 1983).

Teleguide: A Handbook on Video-Teleconferencing, Doug Widner (Public Service Satellite Consortium, 1986).

The Business Television Directory: A Guide to the Corporate Satellite Television Industry, Susan J. Irwin and Michael S. Brown (Irwin Communications, Inc., and Telehealth Associates, 1988).

The Executive Guide to Video Teleconferencing, Ronald J. Bohm and Lee Templeton (Artech House, 1984).

The Information Edge, N. Dean Meyer and Mary E. Boone (Holt, Rinehart and Winston of Canada, Ltd., 1987).

ORGANIZATIONS AND CONFERENCES

ITCA (International Teleconferencing Association), Washington, DC
Telecon, Applied Business Telecommunications, San Ramon, CA
International Television Association (ITVA), Irving, TX
Intelemart, International Teleconferencing Association (ITCA) and Knowledge
  Industry Publications, Inc., White Plains, NY

14

Evaluating Media

By Jeff Kraft
Hyatt Corporation
Denver, Colorado

Until recently, corporate media programs have been evaluated more or less intuitively. Producers would get a feeling that "Yes, the program works for me." Clients would say, "Yes, it correctly includes all of the information I intended." Peers in the media production business would declare, "Yes, it follows our intuitive formula—good production values, comprehensible content, and it is creative." Even the corporate executives would applaud it. But more and more often these days, senior management is demanding greater proof of a media program's contribution to organizational goals.

One big problem with the old methods of relying on "informal feedback" is they have generally neglected the most important media program evaluator—the target audience. And the target audience is the only group which knows for sure whether a program meets its objectives.

DEVELOPING EVALUATION TECHNIQUES

Media producers need to borrow evaluation methods from the social sciences, lessening their reliance on intuition and using more scientific evaluation methods to increase the media profession's credibility. Media professionals should develop a body of supported hypotheses to establish which production techniques work in producing effective programs. Of course, program design formulas won't come easy, but what a useful design tool they would be! It may take additional time and expense to properly evaluate production techniques and final programs, but in the long run, results will be positive.

Since the majority of corporate producers are not trained to evaluate, who will compile this information the field so desperately needs? Video producers will simply have to learn, with assistance from other researchers such as statisticians and trainers. Although most university-sponsored research excludes industrial media, corporate producers can borrow research methods used by academic researchers.

A Continuum

There are various ways to evaluate programs for effectiveness. All are useful, but some are more scientific than others. Quantitative research (assigning numbers to abstract concepts and then using statistics to determine findings) is commonly thought of as more scientific than qualitative research, which relies more on intuition. Based on this principle, an evaluation continuum can be developed. Figure 14.1 shows the range from qualitative to quantitative techniques.

Figure 14.1: Range from qualitative to quantitative techniques

Less scientific	More scientific
Qualitative	Quantitative
Intuition	Randomly administered questionnaires
Focus groups	Controlled experiments
Viewer response cards	Panel studies
Informal feedback	Pre-test, Post-test
	Hypothesis testing
Phone surveys	

All of these methods are useful. Unfortunately quantitative techniques, which provide more valid and reliable results, are more costly and time-consuming—luxuries rarely allowed in today's belt-tightening corporate economy.

No evaluation formula can be applied to every media program. Each program will have a unique evaluation strategy based on its objectives, target audience, budget and distribution. Evaluation design must be based on these factors and planned *simultaneously* with program design. Both must be thought out at the onset of a project. A plan for evaluation must accompany the communication plan.

In fact, evaluation of corporate media programs extends from the first stage of setting objectives (known as formative evaluation) until well after the program's completion (known as summative evaluation).

Both forms require a measuring stick and that is provided by the program's objectives; so clear, measurable objectives are essential to both program success and evaluation.

Consider this example. A client requests a videotape with the following goal: "After viewing the tape, the audience will have a feel for what quality is and why the company needs it." Realistic? Maybe. Measurable? Certainly not. How could you test an audience and quantify how they "feel" what quality is and why the company needs it?

After a rigorous session explaining the need for measurable objectives, we arrive at the following principles: 1) After viewing the program, the audience will be able to define quality; 2) After viewing the program, the audience will be able to list four steps in conforming to requirements; 3) After viewing the program, each audience member will be able to identify his or her product, service, immediate customer, and requirements; 4) After viewing the program, the audience will be able to identify the repercussions of not incorporating quality; and 5) After viewing the program, the audience will be able to state four reasons why quality is important in their work.

These objectives are measurable and work for the program's goal if one assumes that they are a valid measuring stick for "feel for what quality is and why the company needs it." But how will we know if the audience can list four reasons why quality is important in their work? We ask them by way of focus groups, response cards, surveys or whatever other method we determine to be least biased and by whatever methods offer the most valid and reliable results.

The key to beginning an evaluation program is working with clients to establish clear measurable program objectives. From this foundation, evaluation techniques are rather simple to determine.

A Model

Depending on program objectives, different approaches to evaluation may be required. As stated earlier, there is no single evaluation formula that can be applied to every media program. Evaluation can be incorporated, however, into the media production process (see Figure 14.2). This model includes the production steps with which producers are familiar (treatment, script, production, editing) and adds two formative evaluation steps and summative evaluation to the process. The formative evaluation steps guide the producer in getting the program on target, and the summative evaluation suggests to management whether the program meets its objectives, and therefore contributes to company goals.

Let's walk through the process. The client, supported with statistical, historical data, approaches you with a problem that he or she feels will best be solved by a media program. Together the producer and client define the problem and audience in writing and create a goal for the program. Then, the objectives must be written. Objectives must clearly indicate the ability to measure the changes that result from viewing the media package. From the objectives, write a formative and summative evaluation plan. Remember, well-written objectives will determine what is to be measured and will suggest how to measure it.

After the treatment is approved and the rough script is written, conduct a focus group session with 8 to 10 randomly selected members of the target audience. This first focus group should yield useful information in keeping the program on target. The focus group leader should be someone not affiliated with the program, thus reducing bias. The leader should explain the meeting's purpose and how the results of the focus group session will be used. The meeting should be recorded. The leader then should read the rough script aloud.

At this focus group meeting, address three issues:

1. Take note of the audience's general response to the program. (Is the treatment best suited for the target audience? Is the program's message clear?)
2. Determine the ability of the program to meet its objectives. (Can you define quality? Can you identify the repercussions of not using quality?
3. Discuss any production techniques that may be used in the program. (Would this topic be best presented by a male or female spokesperson? Would a vignette help clarify the message?)

The leader should allow everyone to voice opinions and conclude with a question such as, "Can you think of anything we haven't covered that we should have?"

From the information generated in the focus group report, the scriptwriter writes the final script. The program is produced and a rough edit is done on the program. At this point, the second formative evaluation step occurs.

The rough edit should be tested on a random sample of the target audience. This may best be accomplished through a variety of methods, depending on the program. In some cases, a pre-test/post-test method, where increased knowledge should be apparent in the post-test, could be used. In others, a focus group similar to that described earlier could be used. In yet others, a survey questionnaire may be appropriate. At this stage, allow room for improvement in the program by including open-ended questions—ones that encourage the subjects to express what in the program works or doesn't work for them. Keep in mind that this is the first time a sample of the audience has seen the visuals used in the program. All of the information gathered from this second evaluation will be invaluable when making decisions during the final edit.

Now, how can the program fail? The audience has given the producer valuable formative input along the production process. This is where summative evaluation is helpful. Summative research indicates the following: 1) whether objectives are being met; 2) reasons for successes or failures; 3) methods to increase message effectiveness; 4) areas of necessary future research; and 5) need to refine the evaluative process.

A scientific summative evaluation method can be designed with the help of a statistician. Again, use the program's objectives to measure success. Use the combination of methods suggested in the above continuum which most accurately measure the intentions of the program.

If the media package is part of an overall training program, a research design using treatment and control groups could be implemented. A randomly selected treatment group of respondents could go through the training with the media, while another randomly selected control group has the training minus the media. Hopefully, the post-tests on the treatment group will suggest results more in accordance with the program's objectives.

In some cases, evaluation tools may already be in place. For example, a factory experiencing a high rate of parts handling damage produces a training program with the objective of reducing the damage rate. If it meets its objective, data should show a reduction in parts handling damage after employees view the program.

The benefits of these evaluation efforts are obvious. A program producer will have hard evidence concerning the value of his work to the

220 MEDIA FOR BUSINESS

Figure 14.2: Formative Evaluation

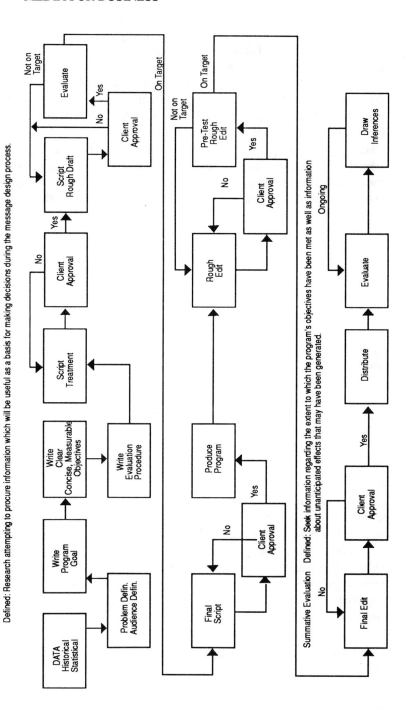

company. If the results are positive, the producer will revel in his success. If they are negative, he will know where to pick up the pieces and work to turn out a better product.

CONCLUSION

Incorporating program evaluation into the production process can be somewhat expensive and time-consuming. It can lengthen the production process by one or two weeks and it requires a commitment from management to make employees available for focus groups, surveys or experiments. It also increases the producer's administrative work load. But without evaluation, producers have no accounting system offering numbers regarding a media program's worth to the company. If the corporate media profession is to advance, programs must be evaluated. A huge step following program evaluation is to build a pool of scientifically supported knowledge suggesting approaches and production techniques that work best in a given situation.

Glossary

A and B roll: A checkerboard method of assembling 16mm camera original into two rolls suitable for producing an answer print. Black opaque leader is alternated with camera original on the A and B rolls, allowing for invisible cuts and lap dissolves.

aerial imagery: A sophisticated and expensive method of transferring multi-image programs to videotape whereby the slide images are projected straight into the video camera.

AFTRA: American Federation of Television and Radio Artists.

AGC: Automatic gain control; a video camera control that automatically adjusts the lens opening based upon the amount of light that is present.

ambiance: See *natural sound*.

animation: Technique whereby the impression of motion is created for film, video and multi-image media presentation. Animation is commonly used to portray an action, operation, or cycle that is difficult or impossible to photograph. Animation is also used when the creative use of this technique will benefit the program.

answer print: The first print from the 16mm A and B rolls. There may be a first, second, third or more answer prints before it is approved.

aspect ratio: The ratio of height to width for various media. Slides have a 2 x 3 aspect ratio; television is 3 x 4; and motion picture film is 4 x 3.

assemble edit: The process of adding new information to a program in sequence.

assembly: The first phase of film editing in which each shot is assembled into its approximate position as indicated by the script.

audio mix: An audio editing session where multiple tracks of audio information are blended onto a single track.

audiographics: A carrier or carrier signal that incorporates both audio and data-based reproducible information (i.e., telephone with fax or with data terminals).

authoring language: A programming language used in interactive videodisc programs. It is more complicated than an authoring system, but provides greater programming flexibility.

authoring system: An easy-to-use authoring tool used in programming interactive videodiscs. It does not require a mastery of computer programming to operate.

branching: Menu options presented to the viewer in an interactive program.

briefing book: A printed copy of information provided to an audience during a media presentation; it is most often used to support slides and overhead transparencies. Briefing books are unique because they duplicate information presented on visuals during the media presentation.

brochure: The Cadillac version of printed material used to support a media presentation. Brochures typically contain a great deal of graphic and photographic information and are commonly printed in color.

camera original: The actual film exposed in the camera during production.

camera-ready art: All elements of a visual pasted onto a piece of artboard and ready to be photographed with no changes in size required.

camera report: A report, filled out at the end of each take of each shot, which lists the scene number, the take number, whether or not sync sound was recorded, camera footage and comments about the shot.

casting director: The person who provides clients with actors and/or models from a variety of talent agencies. The casting director works for a casting agency.

C-band: The most common type of frequency-band satellite, it is subject to interference from local broadcasters.

character generator: A dedicated digital computer that produces letters, numbers, symbols and other graphics information for use in a videotape program.

chargeback: See *cost center*.

codec (coder/decoder): A sophisticated electronic signal processor which enables communication interface with otherwise limited dissimilar systems by narrow-band carriers.

common carrier: An entity which carries signals for anyone who pays. Common carriers are regulated by the FCC and FTC.

compressed audio: Up to ten seconds of voice-grade audio which is stored in a single frame of a videodisc.

computer graphics: Graphics created with the aid of a specially designed computer.

conformer: A film professional trained to handle and conform original camera footage.

conforming: The act of cutting the camera original to match the edited workprint.

continuous tone film: Black-white film that records a full range of light and dark tonal values.

contingency: A percent of the total budget (usually 10% to 15%) used to compensate for any extra costs not included in the budget.

copy stand: A unit equipped with lights and a stand on which to mount a camera for still, video or motion picture photography.

corporate teleplay: A script format in which visuals are described in text form, running from left to right margins. Visual information includes interior or exterior shot selection, scene location, time of day and a description of the action. Music, sound effects, natural sounds and on-camera dialog are centered on the page. Voice-over narration is capitalized and placed in the right third of the page.

cost center: A corporate accounting system in which an in-house media production department charges all production costs back to the in-house customer. The media production department functions as a business within a business.

cut: A direct video transition from one shot or scene to the next.

demographics: Factors that help to define or categorize a group of people, such as age, sex, educational background and marital status.

depth of field: The distance from the closest to the furthest point in/a visual (slide, photo, video or film image) that is sharply focused.

desktop publishing: A method of producing quality text and graphics information using a personal computer, specialized software, and an output device such as a laser printer.

digital effects: Video transitions accomplished with the use of a switcher and other equipment which do the following: 1) reduce or expand an image shape and/or size; 2) re-position the image; 3) change the image's aspect ratio; 4) flip, tumble or crop the image. Other effects include star trails, blinks, posterizations and mosaics.

directionality: The pattern by which a microphone picks up sound.

dissolve: A visual transition whereby two images temporarily overlap as one fades in and the other fades out.

dissolve unit: A device that controls the rate at which slide projector lamps turn on and off.

dolly: A wheeled platform on which a camera or tripod is mounted for smooth camera movement.

downlink: A facility that receives signals directly from a satellite.

dubber: A machine that is basically a tape recorder; it records and re-records audio information.

edge number: An alphanumeric image recorded on the edge of 16mm film at 20-frame intervals. Edge numbers are used to pinpoint the location of individual shots on motion picture film.

edit controller: A piece of video equipment that is connected to at least one video player and one video recorder to allow the editor to perform the specific transition that links one shot to another.

establishing (master) shot: A single shot in a video or motion picture production that incorporates all elements of the larger scene. The establishing shot is usually the first shot photographed prior to close-ups and reverse angles.

fade: A transition in which a visual fades to black or a black screen fades to a visual. Termed "fade in" or "fade out."

fiber optics: Allows the transmission of signals through glass-strand cable with no degradation of signal.

film stock: Raw, unexposed film.

fine cut: The final phase of film editing in which the rough cut is honed and polished until the film is complete.

flatbed editor: A newer machine used for editing motion picture film and audiotape in synchronization. Reels of film and tape are mounted horizontally on the editor.

flowchart: A chart created in the pre-production phase of the interactive process that diagrams every possible option the program offers.

focus group: A group of randomly selected individuals who are representative of a program's target audience and who voice opinions about specific topics or programs.

font: All characters within a given typeface, including upper case, lower case and numbers.

formative evaluation: Evaluation which takes place before the program's completion.

full coat: 16mm film stock coated with oxide and used for audio recording.

gateway: A device that allows dissimilar systems to interconnect.

glass mount: A type of slide mount used instead of a plastic or cardboard mount. These mounts keep the slides clean and are less likely to jam in the slide projector.

graphic artist: The person responsible for designing and producing graphics for media presentations.

graphics: Those images that consist of titles and credits, words, illustrations, charts, graphs, tables, figures and other written information.

handouts: One or several pages of information provided to the audience before or after a presentation to support it. Typically, handouts are typewritten or typeset and feature little or no artwork.

hard copy: A printed copy of information presented during a media presentation.

hybrid system: A corporate accounting system incorporating elements of both chargeback and overhead systems.

illustration: A drawing or representation of an object, concept or action.

insert edit: Any edit in which video or audio information is inserted over existing information.

instructional designer: A person who helps establish goals, learning objectives and tasks in an educational or training program. This person also helps determine the learning strategy.

Ku band: A type of frequency-band satellite. It is more expensive than C-band, but uses smaller hardware and is less likely to be subject to local broadcast interference.

lavaliere: A small microphone with an omnidirectional pickup pattern. It is also called a "tie clip" microphone.

Level I: Refers to an interactive system that consists of a television monitor, a keypad and a videodisc player.

Level II: Refers to a more sophisticated system which contains all the components found in Level I, in addition to a memory, the ability to accept a short computer program and the capacity to store information.

Level III: The most advanced type of interactive system, it uses an external computer to control the videodisc and accepts input from the user.

linkages: Types of interconnected distribution used in electronic responsive-means communication. Examples include common carriers, low-power broadcast, fiber optics, wave form, microwave, cable, satellite and telephone lines.

log sheet: A report produced from the workprint which lists edge numbers, scene numbers, take numbers and scene descriptions. The log sheet is made just prior to beginning the edit.

mastering: The process of collecting audio and video information onto a videotape which, in turn, will be made into a videodisc.

media producer: The individual responsible for coordinating all phases of media production.

microwave: A type of transmission system that is capable of sending signals for distances up to 150 miles.

mixer: An audio device that combines sound from different sources and sends that sound to various outputs.

moviola: An older machine for editing motion picture film and audiotape in synchronization. Reels of film and tape are mounted vertically on the moviola.

multi-image: Multiple slide images which are programmed to a pre-recorded soundtrack and projected onto a screen.

multi-media: The use of several media in one presentation or setting.

music library: A collection of commercial music licensed by a company from which media producers may purchase specific selections. Music libraries usually offer a variety of musical styles and playback options (i.e., CD, LP, cassette, reel-to-reel).

natural sound: Extraneous audio that is recorded when shooting videotape or motion picture film. Examples of natural sound are wind, traffic noises and birds chirping.

needle drop: A fixed fee paid by media producers to a music library for use of a specific music selection(s).

negative imaging film: Film which, when exposed, creates a negative image with colors and light and dark values reversed. Negative imaging films are used to make positive prints or transparencies.

objectives (measurable): Clearly written, specific purposes that measure how and what an audience feels about a program after viewing it.

operating budget: The budget allocated to an in-house media production department to cover all costs for a one year period.

optical track: A high-contrast picture of the soundtrack on 16mm films. The optical track runs along the side of the film where the sprocket holes would normally appear.

original artwork: Any artwork, illustrations, charts, graphs, cartoons, etc., created for a specific presentation.

overhead system: A corporate accounting system in which the in-house media production department is provided with a budget to cover all production costs at no cost to the client department.

pamphlet: A folded or, in some cases, bound package of information provided to an audience before or after a presentation. May include charts, graphs or other types of supporting graphic elements.

paste up: The process of creating camera-ready artwork by pasting the graphic elements of a visual (copy, fine art, illustrations, etc.) into their proper positions on a piece of artboard.

PBX (private branch exchange): An in-house, locally accessed telephone network which provides better quality and greater simultaneous capabilities than do regular local network telephone carrier systems.

pilot: A pre-release of the interactive program which allows the program to be evaluated and revised before mastering onto videodisc occurs.

positive imaging film: Film which, when exposed, creates a positive image with a normal range of color or light and dark values.

positive print: A film image that, when projected, depicts a full range of colors as they normally appear (as opposed to a negative image which reverses natural colors when projected).

post-production: The final phase of a media production. It may include editing, programming, pulsing or any other process whereby the visual and sound elements are combined to form a final program.

pre-production: The planning stage of any media production.

presentation analysis: A structured approach to collecting information about a media production prior to beginning the production process.

primary audience: The group of people specifically targeted to view a media program.

production: In simplest terms, the process of producing visuals and sound for a media production.

program budget: The amount of money allocated for production of a single media production.

programmer: A piece of equipment, usually a dedicated computer or a computer board in a PC. When used in conjunction with a dissolve unit, it controls the rate at which slide projector lamps turn on and off. Programmers also control the forward or backward movement of the individual slide trays.

quantitative research: Assigning numbers to abstract concepts and then using statistics to determine findings.

random access: A videodisc function that allows the viewer to go to any part of the program within three to five seconds.

rear projection: A method of showing multi-image programs whereby the program and equipment are situated behind a rear-projection screen.

release print: The final print of a completed film which is released to the customer and shown to the audience.

rental costs: Anything rented in support of a media production. Examples include: cameras, lighting equipment, audio equipment, and studio facilities.

rough cut: The second phase of film editing in which each shot is trimmed and adjusted. Creative decisions are made which affect the film's ultimate timing and pace.

rough draft: A detailed written description of a media program's visuals and audio content.

SAG: Screen Actors Guild.

satellite: A type of transmission system that orbits the earth above the equator and is capable of transferring signals to downlinks while reproducing and distributing signals from uplinks.

script format: Generally, four types of formats are available to writers and each has its own advantages and disadvantages. Formats include the split page, storyboard, corporate teleplay and Writer's Guild of America style.

secondary audience: An identifiable group of people (other than the primary audience) who are likely to watch a media program.

shooting ratio: The number of slides, videotapes or feet of film shot compared to the number used in the program.

shot list: A detailed description of each visual that will be needed in the final program.

shotgun: A microphone with a highly directional pickup pattern; best for capturing sound within a narrow forward angle over relatively long distances.

Glossary 233

soundtrack: The single track of audio information produced during the audio mix. The soundtrack includes all dialog, music, ambient sound and sound effects that will be heard in the final product.

special effects: Visual transitions that change one shot or scene to another through optical effects. Examples include wipes (patterns and geometric shapes), digital effects (page turns, tumbles, etc.) and squeeze zooms.

special effects generator: A device that processes several video signals in order to produce a special effect such as a wipe or a pattern.

split page: A script format in which visual descriptions appear in the left column of the page and audio descriptions appear in the right.

spoken word: Voice-over narration or on-camera monolog/dialog.

staging: The process of determining the logistics for the performance of a multi-image program. This includes location scouting, set-up and tear down of equipment, and showing of the multi-image program.

stock photos or footage: Photos, slides, motion picture film or videotape purchased from "stock houses" by media producers when the needed visuals cannot otherwise be obtained. Examples include seasonal shots, historical visuals, images of foreign lands, generic industrial shots, etc.

storyboard: A script format in which pictures depict the visuals. Beneath the drawings appear notes concerning camera movements, transitions and audio. Storyboards are completed and submitted to the client for approval before production begins.

summative evaluation: Evaluation that takes place after a program's completion.

switcher: The input control console used to select or mix the video output.

sync (synchronization): An audible or inaudible pulse recorded on the soundtrack of a slide/cassette program to indicate when a slide change should occur.

talent: Any person, professional or amateur, who appears on-camera or who does narration.

talent agency: A business that makes available actors and/or models for audio and video productions.

technical expert: A person who provides subject matter and technical expertise to the scriptwriter, the producer and the director of a media program.

teleconference: An umbrella term used to describe meetings or presentations whereby geographically scattered persons are able to communicate via audio and/or video means.

teleprompter: A device that fits near the lens of a video or film camera and scrolls the script for the on-camera talent.

text-heavy: Term used to describe a media presentation relying primarily on visuals consisting of text or word charts.

tilt: The up and/or down movement of the camera head.

timer: The person responsible for adjusting the color filtration and exposure when printing an answer print from the A and B rolls.

track: The area of the tape occupied by a recorded signal; denoted as two-track, four-track, etc.

transitions: The visual techniques employed to move from one scene to the next. The most common transitions are the cut, dissolve, fade in/out and special effects. Transitions are achieved in the editing or programming phase of a program.

transparency: A transparent photographic image which can be projected onto a screen for use during a media presentation. Transparencies can be black-and-white or color and are most often 35mm slides, overhead transparencies or motion picture film.

treatment: A written narrative that describes what the audience will see and hear in a media program.

truck: Lateral movement of the camera and its support to the left or right of the subject.

TV safe area: The rectangular area in the center of a film frame where titles can be placed and safely transferred to videotape.

type size: The size of the letters chosen for the camera-ready artwork. Type height is measured in points, with one point equaling 1/72".

typeface: The style or design of type.

typesetter: The person who produces copy (in the size and font indicated by the graphic artist) which is pasted onto the camera-ready artwork for media presentations.

uplink: A facility or device that can send signals to a satellite for further distribution.

videoconference: A multiple-party meeting wherein participants communicate through audio and video signals transmitted by cable, microwave, satellite or other means.

videodisc: A piece of plastic about the size of a record album that contains visual and/or audio information. Its primary use is in interactive programming.

white balance: The video camera adjustment that produces correct color balance. This is done when the camera is aimed and focused on a flat white field such as a piece of cardboard, paper or even a white shirt.

wild sound: Ambient noise recorded without a picture.

wireless microphone: One that transmits a signal to a receiver on a specific frequency.

workprint: A positive film print made from the camera original, which is physically cut during the editing session. The workprint is not color corrected.

zoom: A change in the focal length of the camera lens so that the subject appears closer (zoom in) or farther away (zoom out).

Bibliography

Alten, Stanley R. *Audio in Media.* Second edition. Belmont, CA: Wadsworth Publishing Company, 1986.

Bishop, Ann. *Slides—Planning and Producing Slide Programs.* Rochester, NY: Eastman Kodak Company, 1984.

Blum, Richard A. *Television Writing from Concept to Contract.* Revised edition. Boston and London: Focal Press, 1984.

Blumenthal, Howard J. *Television Producing and Directing.* New York: Barnes & Noble Books, 1987.

Brandt, Richard. *Videodisc Training: A Cost Analysis.* Falls Church, VA: Future Systems Inc., 1987.

Bunyan, John A. *Why Video Works.* White Plains, NY: Knowledge Industry Publications, Inc., 1987.

Burrows, Thomas D. et. al. *Television Production Disciplines and Techniques.* Fourth edition. Dubuque, IA: William C. Brown Publishers, 1989.

Cartwright, Steven R. *Training with Video.* White Plains, NY: Knowledge Industry Publications, Inc., 1986.

Cartwright, Steve R. *Secrets of Successful Video Training: The Training with Video Casebook.* New York: Knowledge Industry Publications, Inc., 1990.

Chamness, Danford. *The Hollywood Guide to Film Budgeting and Script Breakdown.* Revised edition. Los Angeles: The Stanley J. Brooks Company, 1977.

Compesi, Ronald J. and Ronald E. Sherriffs. *Small Format Television Production*. Newton, MA: Allyn and Bacon, Inc., 1985.

Crowell, Peter. *Authoring Systems: A Guide for Interactive Videodisc Authors*. London: Meckler, 1988.

Edmonds, Robert. *Scriptwriting for the Audio-Visual Media*. Second edition. New York: Teachers College Press, 1984.

Elser, Art. "An Introduction to Interactive Videodisc," *Technical Communication*, 35, no. 3 (third quarter 1988).

Feininger, Andreas. *The Complete Photographer*. Englewood Cliffs, NJ: Prentice-Hall, Inc., 1965.

Floyd, Steve and Beth, ed. *Handbook of Interactive Video*. White Plains, NY: Knowledge Industry Publications, Inc., 1982.

Freeman, Michael. *The 35mm Handbook*. London: Quarto Publishing, 1980.

Gordon, Roger L. *The Art of Multi-Image*. Albington, PA: Association for Multi-Image International, 1983.

Gross, Lynne Schafer. *The New Television Technologies*. Second edition. Dubuque, IA: William C. Brown Publishers, 1983.

Hedgecoe, John. *The Photographer's Handbook*. New York: Alfred A. Knopf, 1978.

Heinich, Robert et. al. *Instructional Media*. Second edition. New York: Macmillan Publishing Company, 1986.

Holmes, Nigel. *Designers Guide to Creating Charts and Graphs*. New York: Watson-Guptill Publications, 1984.

Iuppa, Nicholas V. *A Practical Guide to Interactive Design*. White Plains, NY: Knowledge Industry Publications, Inc., 1984.

Kemp, Jerrold E. *Planning and Producing Instructional Media*. Fifth edition. New York: Harper and Row, 1985.

Kenny, Michael and Raymond F. Schmitt. *Images, Images, Images—The Book of Programmed Multi-Image Production*. Third edition. Rochester, NY: Eastman Kodak Company, 1983.

Kindem, Gorham A. *The Moving Image*. Glenview, IL: Scott, Foresman and Company, 1987.

——————. *Kodak Guide to 35mm Photography*. Second edition. Rochester, NY: Eastman Kodak Company, 1984.

Lambert, Steve and Jane Sallis, eds. *CD-I and Interactive Videodisc Technology*. Indianapolis, IN: Howard W. Sams and Company, 1987.

Lipton, Lenny. *Independent Film Making*. Revised edition. New York: Simon and Schuster, 1972.

Mager, Robert F. *Measuring Instructional Intent*. Belmont, CA: Fearon Publishers, 1973.

Malkiewicz, Kris J. *Cinematography: A Guide for Film Makers and Film Teachers*. Revised edition. New York: Van Nostrand Reinhold Company, 1973.

Marlow, Eugene. *Managing Corporate Media*. Second edition. White Plains, NY: Knowledge Industry Publications, Inc., 1989.

Matrazzo, Donna. *The Corporate Scritpwriting Book*. Philadelphia: Media Concepts Press, 1980.

McQuillin, Lon. *The Video Production Guide*. Indianapolis, IN: Howard W. Sams & Co., 1983.

Mezey, Phiz. *Multi-Image Design and Production*. Boston and London: Focal Press, 1988.

Millerson, Gerald. *Effective TV Production*. Second edition. London: Focal Press, 1983.

Oakey, Virginia. *Dictionary of Film and Television Terms*. New York: Harper and Row, 1983.

Reisz, Karel and Gavin Millar. *The Technique of Film Editing*. New York: Hastings House, 1968.

Roberts, Kenneth H. and Win Sharples, Jr. *A Primer for Film-Making*. Indianapolis, IN: The Bobbs-Merrill Company, Inc., 1978.

Schmid, William T. *Media Center Management*. New York: Hastings House, 1980.

Schneider, Eward W. and Junius L. Bennion. *Videodiscs*. Englewood Cliffs, NJ: Educational Technology Publications, Inc., 1981.

Schwier, Richard. *Interactive Video*. Englewood Cliffs, NJ: Educational Technology Publications, Inc., 1986.

Sigel, Efrem et al. *Video Discs—The Technology, the Applications and the Future*. White Plains, NY: Knowledge Industry Publications, Inc., 1980.

Swain, Dwight V. *Scripting for Video and Audiovisual Media*. London and Boston: Focal Press, 1981.

Upton, Barbara and John. *Photography*. Boston: Little, Brown and Company, 1976.

Van Deusen, Richard E. *Practical AV/Video Budgeting*. White Plains, NY: Knowledge Industry Publications, Inc., 1984.

Van Nostran, William. *The Scriptwriter's Handbook*. White Plains, NY: Knowledge Industry Publications, Inc., 1989.

Wiese, Michael. *Film and Video Budgets*. Stoneham, MA: Focal Press, 1984.

Wurtzel, Alan and Stephen R. Acker. *Television Production*. Third edition. New York: McGraw-Hill Book Company, 1989.

Zettl, Herbert. *Television Production Handbook*. Fourth edition. Belmont, CA: Wadsworth Publishing Company, 1984.

About the Authors

Robert H. Amend is an Associate Professor of Technical Communication at Metropolitan State College in Denver, CO. He has written, produced and directed media programs for more than twelve years. He serves as an industry consultant and is co-founder of ScriptRight Services, Inc. He has received awards for his writing and video production from the International Television Association (ITVA).

Michael A. Schrader is a corporate television producer for EG&G Rocky Flats in Golden, CO. He graduated from California State University at Northridge with a degree in television and film production. He has been active in corporate communications for over twelve years and previously served as manager of Media Services for Rockwell International. He has received awards for his video productions from the International Television Association (ITVA).